Ground Truth

Ground Truth
The Future of U.S. Land Power

Thomas Donnelly and Frederick W. Kagan

The AEI Press

Publisher for the American Enterprise Institute

WASHINGTON, D.C.

To the men and women of the
United States Armed Services

Distributed to the Trade by National Book Network, 15200 NBN Way, Blue Ridge Summit, PA 17214. To order call toll free 1-800-462-6420 or 1-717-794-3800. For all other inquiries please contact the AEI Press, 1150 Seventeenth Street, N.W., Washington, D.C. 20036 or call 1-800-862-5801.

Library of Congress Cataloging-in-Publication Data

Donnelly, Thomas, 1953–
  Ground truth : the future of U.S. land power / Thomas Donnelly and Frederick W. Kagan.
    p. cm.
Includes bibliographical references.
    ISBN-13: 978-0-8447-4262-5 (pbk.)
    ISBN-10: 0-8447-4262-7
    1. United States—Armed Forces. 2. United States—Military policy. 3.United States. Army. 4. Military readiness—United States. 5. Military planning—United States. 6. Unified operations (Military science)—United States. I. Kagan, Frederick W., 1970- II. Title.

    UA23.D6226 2008
    355'.033273—dc22

                                                      2008013062

    12 11 10 09 08   1 2 3 4 5

Contents

Introduction

In wartime Washington, the one point of bipartisan agreement is that our land forces are too small. Indeed, the small size of America's land forces has for years been the tightest constraint on U.S. military strategy, and it is likely to remain so for years to come. The failure to foresee the need for larger ground forces, and the reluctance to respond rapidly once the requirement became undeniable, may constitute the most profound mistake made by the Bush administration after the attacks of September 11, 2001. It will fall to the next administration to begin the effort to remedy this most dire divergence between American military ends and American military means, and the undertaking is likely to go on for at least a decade. We went to war with the army we had, and now must accept the truth about the size, shape, and costs of the land forces we need.

This book is an attempt to tell that truth. It addresses the most important policy questions about U.S. land power—questions that must be answered if we are to create the forces we need. It looks at the state of the current force, and at the missions that American land forces must be prepared to conduct. It seeks to consider the nature of land warfare in the early twenty-first century, and to infer, in turn, the qualities American forces will need for success on these new battlefields. And it provides specific answers to the questions of how large the force should be, how it should be equipped and structured, and what it will cost.[†]

---

† We would like to thank the dedicated research assistants and interns of the American Enterprise Institute, who have been instrumental in bringing this project to fruition.

1

Chapter 1, which deals with the missions confronting U.S. land forces, begins by charting the slow and painful reduction of these forces since the end of the Cold War. We suggest that the desire for a peace dividend, coupled with the belief that modern information technology would transform the nature of war, led the United States to view land war as a thing of the past. But the United States is coming to terms with the fact that it cannot simply fight the way it would prefer to fight—rapidly, decisively, and from a distance. The land force, in our post-9/11 and "post-Baghdad" world, is deluged with multiple missions, and we indicate in this chapter how the military ought to prioritize them, identifying enemies, threats, and challenges. We acknowledge that the many commitments of the Long War will be the main preoccupation of the active-duty U.S. land forces, but we also consider a number of additional scenarios that will affect the size and structure of the army and marines, ranging from a nuclear Iran to state collapse in North Korea, Pakistan, or Saudi Arabia.

If the number and duration of operations since 9/11 have been unexpected, so too has the nature of these wars. In chapter 2 we look at the types of conflicts the United States must be prepared to engage in. Since even the most experienced strategists err in predicting what kind of war will be next—and since the army cannot speedily adapt itself to each new conflict—the United States cannot simply prepare for one kind of expected combat (for example, "asymmetric"). It needs to be ready to fight across the entire spectrum of conflict, from high-end major war to low-end peacekeeping mission. It needs, moreover, to be prepared to undertake these fights alone. Even if our allies are willing in spirit, they tend to be weak in flesh, having disarmed even more significantly than we have. Because the heaviest burden of current and likely future threats will fall on U.S. ground troops, we argue, the army and Marine Corps must be rebuilt before a new crisis engulfs us.

Chapter 3, a series of short case studies of recent campaigns, is designed to highlight the emerging battlefield realities and to offer lessons to military planners. Given the pace at which U.S. ground forces are moving, these studies can only be suggestive—of the range, the length, and the intensity of modern land force combat and stability operations; this is a series of snapshots. Nevertheless, they give a better appreciation of the details of the missions faced by U.S. land forces. They also offer a range of important lessons for the future—from the proper ratio of troops to population in counterinsurgency operations, to the likely effect of years of reliance on airpower on the performance of ground troops, to the value of cultural training and of initiative by junior leaders.

The fourth chapter is an attempt to distill the essential qualities that U.S. land forces will need in order to achieve success in their current and imminent missions. Given the number and variety of missions and the emerging nature of land war, it is apparent that U.S. land forces need not only to be more numerous but also to reflect capabilities beyond the timely and devastating delivery of firepower—though the need to deliver firepower undoubtedly remains. We argue that our troops need to be forward stationed: good things tend to happen when U.S. forces are present, and bad things happen when they are not. We argue further that they need, in this war of information, to be able to effectively gather, share, analyze, and respond to a flood of information; that better and more extensive education and training for leaders, particularly junior leaders, is needed; and that the land forces must be able to build effective partnerships with the foreign militaries they help to develop and train.

In the final chapter, we answer some of the toughest questions about the increased land force, including precisely how big the force should be and what it will cost. Breaking down the various forces required for a variety of possible commitments, we argue for a total active-duty land force of about one million, with a reserve component of approximately 850,000 to 900,000, an army reserve of about 350,000, an Army National Guard of 450,000 to 500,000, and a marine reserve of up to 50,000. The costs of building such a force are undoubtedly large—for the army, with an active-duty force of 800,000, around $240 billion—but the proper question to ask is not whether we can afford sufficient land forces, but whether we will choose to have them. In the context of the overall American gross domestic product, the total burden of an expanded army, by the time this rebuilding process neared completion in around 2017, would be less than 1.2 percent of projected GDP.

There now exists a wealth of knowledge about what U.S. land forces need to be and to do, in spite of the dizzying diversity of missions they face and the difficulty of victory. There thus exists a solid basis for future force planning. The solution is no longer, to use a term popularized by former defense secretary Donald Rumsfeld, an "unknown unknown." This study offers a specific way forward, with the recognition that building the required force is likely the effort of a decade. The "ground truth" we offer may be imperfect or incomplete, and it is certainly likely to provoke contention. We advance it as a basis for discussion but also as a plan of policy action—action that is overdue and urgent.

1

The Mission

American military planners had it easy during the Cold War in one respect at least: they knew what battle they had to prepare for. From 1945 to 1989, everyone expected that the decisive battle of World War III would be in Europe, if it ever came. America's soldiers could often identify the precise locations where they expected to fight. The enemy's order of battle was known, the technical capabilities of his systems understood, and his tactics and doctrine extensively studied. American warriors cycling through training centers in the California desert practiced against "Krasnovians," the forces of a fictitious country that fought in the standard formations of the Red Army with weapons simulating T-72 tanks, MiGs, and Hind helicopters. These assumptions about a future conflict's location and nature might have been quite wrong, of course. The navy worked hard in the 1970s and 1980s to remind America's leaders that the United States had critical interests in the Pacific and around the world that it was not prepared to defend. And if World War III had gone nuclear at once, then the massed armies on the central front would likely have proved irrelevant. But the consensus, right or wrong, about the nature and location of the key threat made defense planning as easy as it can ever be.

Nothing is easy about planning America's defenses now. There is no consensus about the nature of the threat or its location. There is, however, general agreement that our most dangerous enemies—terrorists—have no fixed order of battle and use weapons like box cutters and civilian airliners whose technical specifications are unimportant. Is Iraq the "central front" in this war? Some question if it is any part of the war on terror at all. Some have even questioned whether there is, can, or should be a war on terror. Few understand what such a war would be.

Beyond the war on terror, there is no agreement at all about the nature and scale of the threats America faces. Most politicians deplore North Korea's acquisition of nuclear weapons, but virtually none advocate doing anything about it beyond getting the Chinese to talk to Pyongyang. Most leaders also warn against the dangers of a nuclear Iran, but the preferred option for almost all—including, apparently, the Bush administration—is talk and sanctions rather than combat. There are understandable reasons for the desire to reject or ignore military options in these scenarios—intervention in Korea today runs the same risk of escalation with a powerful China that it ran in the 1950s; the difficulties American arms have encountered in Iraq seem even more daunting when scaled up for the size and complexity of Iran. And adherents of the postmodern view of war as an atavistic, outdated, barbaric, and ineffective way to resolve conflicts take visible satisfaction in the checks the American military has encountered in Iraq. Diplomacy, they say, is the only way to resolve crises in today's world, where military intervention always and inevitably makes things worse.

How We Got Here

America has faced such periods of confusion before, but never when the stakes were so high. As the U.S. Army focused on defeating American Indians at the end of the nineteenth century, its leaders thought about preparing for wars in Europe that no one expected the United States to become involved in. After the partially successful conclusion of precisely such an intervention in 1918, America's leaders once again lost touch with their nation's core interests. The desire to avoid further international entanglements, to focus on domestic priorities at a time of profound domestic crisis, and to turn away from a world with which many Americans were disenchanted overwhelmed the counsels of reason, and America disarmed its ground forces almost completely. The folly of that decision was made manifest on December 7, 1941, and shortly thereafter when Hitler declared war on the United States. Having failed utterly to keep pace with the real requirements for securing America's growing interests in Europe and the Pacific after World War I, the United States had to race to build a military almost from scratch. The effort was successful, but the price was high—far higher than that of maintaining a military commensurate with America's power and the requirements of its own defense in the interwar years.

The United States repeated this pattern after World War II— disarming so thoroughly and rapidly that the country had to race to rearm not five years after the war had ended to fight a much more limited contingency in Korea. Thereafter, the exigencies of the Cold War and the persistence of a clear existential threat in the Soviet bloc helped ensure that the American military never reached a point at which it could not defend America's most vital interests, although there were times when America would have had to abandon many secondary interests and even when success on the fronts expected to be decisive was in doubt. But the Soviets were ever a cautious and predictable foe. The balance of nuclear deterrence worked reasonably well against a regime concerned primarily with its own survival and only secondarily with advancing its agenda around the world. And America's powerful nuclear arsenal covered numerous deficiencies in conventional forces as well. There were many years in which the Soviets could reasonably have expected to overrun Europe with conventional arms if they had chosen, but none when they could have been confident of deterring or winning the nuclear war likely to result. And that danger was enough to prevent them from ever trying.

Confusion grew once again as the Cold War ran down, and the 1990s saw a proliferation of theories about the future of conflict, the likely nature of future war, and the identity and characteristics of enemies in struggles placed comfortably into the future. The decade was widely thought to be a "strategic pause," a time when no war was likely for a protracted period of time; a time when "peace dividends" could be distributed at the expense of military capability and readiness; a time when the world could be allowed to pursue its own course with little American involvement. The army began to plan not just for the transformation of its force under slogans like "Force XXI," but even for more futuristic endeavors like "Army After Next." The air force considered "Air Force 2025" in the 1990s. The Joint Chiefs of Staff produced "Joint Vision 2010" and then "Joint Vision 2020" (when it became apparent that 2010 was rather too soon for all the transformations envisioned). The navy became enamored of advanced communications and developed concepts like "network-centric warfare" based on the supposed lessons of the information revolution in the business world. Very few of these theories identified any particular enemy against whom the forces they proposed—or demanded—would fight. Very few took account of the real conflicts taking place as they were being developed, described, and advocated.[1]

For the 1990s was not really a strategic pause. On the contrary, it was one of the busiest decades the American military has ever had. Over six hundred thousand American troops—as many as served in Vietnam at the height of our deployment there—drove Saddam Hussein from Kuwait in 1991. A much smaller force was driven out of Somalia in humiliation in 1993. Another force was deployed to Haiti in 1994 expecting to fight, but fortunately landed unopposed. Combat was required the next year to force warring parties in Bosnia to accept nearly fifty thousand peacekeepers. Additional combat was required to permit the insertion of a smaller peacekeeping force into Kosovo in 1999. A war scare in Kuwait sent 150,000 soldiers scurrying to the Persian Gulf in 1994, as another war scare on the Korean peninsula forced President Clinton to consider how many casualties he was willing to risk to prevent Pyongyang from developing its nuclear program. Saddam's continued cat-and-mouse game with inspectors led to repeated troop movements and ultimately to a large-scale missile strike in 1998. American warplanes patrolled the skies over northern and southern Iraq continuously from 1991 through 2003 (and continue to fly over all of Iraq today). The continued movement and engagement of tens and sometimes hundreds of thousands of American troops should have alerted America's leaders to the weakness of the assumptions underlying the strategic pause concept, and should have called into question military planners' focus on the distant future, especially their widely trumpeted claim that ground forces would be irrelevant in future conflict.[2]

But the attractions of cheap and sterile war offered by the advocates of airpower solutions and technologists overcame the realities on the ground. The peace dividend was distributed in a bipartisan way, by both the first Bush and the Clinton administrations under both Democratic and Republican Congresses. The military was reduced by around 40 percent across the board, and programs promising high-tech, long-range precision-strike capabilities were preferred to more traditional boots-on-the-ground.

Worse still, the military changed the description of its mission in a deliberate effort to avoid involvement in peacekeeping efforts. By the middle of the decade, senior army officers described the army's primary purpose as "fighting and winning the nation's wars," relegating "operations other than war" (including peacekeeping and stabilization operations) to secondary status.[3] The change was not noticeable on the ground, as the American military spent most of the 1990s engaged in precisely such low-intensity and highly political undertakings. Nor was it reflected in the Clinton administration's strategy documents, which highlighted the importance of operations other than war.[4] But defining such efforts as "lesser included missions" under the primary mission of "fight and win wars" was an important manifestation of the military's preferences and priorities. Those priorities carried into the George W. Bush administration, which invented very little of its own in the military realm, but adopted many of the most unrealistic and out-of-touch visions that had developed in the Clinton years.

Even as a presidential candidate, Bush advocated high-tech, long-range airpower solutions and denigrated peacekeeping and nation-building activities. As president, he worked through Secretary of Defense Donald Rumsfeld to impose this vision on the military. Rumsfeld, it was widely rumored at the start of his tenure, wanted to cut the remaining active army divisions by another 40 percent, which would have amounted to around a 75 percent reduction from Cold War levels. Debates about refocusing

resources away from ground forces and into airpower and information technology were reaching their apotheosis as al Qaeda operatives flew civilian airliners into the World Trade Center, the Pentagon,
and—thanks to the incredible courage of the passengers—a field in
Pennsylvania rather than the White House or the Capitol. The
attacks ended the discussion of reducing the ground forces, but
they did not bring clarity to our understanding of a world that suddenly seemed inexplicably and unpredictably dangerous a day after
it had seemed obviously and predictably safe.[5]

In the context, there is nothing surprising in the confusion
that characterized defense policy discussions in the wake of the
9/11 attacks. Although al Qaeda first struck the American homeland in 1993 and struck U.S. outposts around the world repeatedly
over the next eight years—earning a well-deserved if ineffective
missile strike against its bases in Afghanistan in 1998—few people
knew anything about the organization and almost no one knew
how extensive, dangerous, or effective it could be. The whole development of militant Islamism from the 1960s on, in fact, had largely
passed Americans by. Apart from a handful of experts, virtually
no one had recognized the increasingly organized, militant, and
sophisticated ideologies and movements that guided radical Muslim
terrorist organizations from the 1970s on. Ayatollah Khomeini's
seizure of power—and American hostages—in 1979 seemed to
most people to be an outlier. Apart from specialists, few people
even within the military noticed that most of America's military
operations in the 1980s were directed against terrorists only loosely
related to Soviet sponsors, or that our operations in the 1990s
continually involved us in struggles within the Muslim world,
whether in Africa (Somalia), the Middle East (Iraq and Kuwait),
Central Asia (Afghanistan and Pakistan), or Europe (Bosnia and
Kosovo). The geopolitical rationale of the Cold War required placing all military activities within the context of the struggle against
Communism, a tendency that obscured the development of
Muslim extremism and state-sponsored terrorism.

The belief in a strategic pause and the focus on the question of
peacekeeping or not peacekeeping, as well as the geographical

dispersion of American military interventions, prevented a clear recognition of the dawning challenge in the 1990s. Virtually no Americans, after all, think of Europe as part of the Muslim world; many did not even see Africa in that way. And our interventions in the Balkans and Somalia were not intended to involve us in religious or sectarian conflicts with Islam, even though they did.

But if the confusion surrounding American strategy and defense policy in the immediate aftermath of the 9/11 attacks was understandable, the continuing confusion that clouds the debate today is less so. Over the past six years, much has become clear about the nature of the current enemy, the scale of the current threat, successful and unsuccessful means of fighting it, and where the decisive theaters in this conflict are likely to be. The challenges of nuclear proliferation, especially within the Muslim world, have also become clearer than they ever were before, and the challenge posed by the most dangerous regime in that world—Iran—is easier to understand. It is quite possible now to articulate far more clearly and definitively than has been the case for some time the framework of the missions America's armed forces must be able to execute.

The Military's Missions

The confusion over American strategy and defense policy in the 1990s bred a proliferation of increasingly complex statements of the missions for which the military should be designed and maintained. The primary mission that supposedly shaped the size and organization of the military in the 1990s, flowing from the premise that the requirement was to "fight and win the nation's wars," was the ability to fight two nearly simultaneous "major regional conflicts," or MRCs. MRCs were variously defined, but the requirement to fight two always added up, curiously enough, to a military sized and shaped as it was in projections and plans made in 1992 and 1993—throughout the decade and through a bewildering series of changes in military operational patterns and the international setting.[6]

The Bush administration decided in 2001 that this two-MRC framework was too restrictive and did not take sufficient account of the complexity of the challenges facing the nation, particularly after 9/11. It replaced that construct with an even vaguer and more perplexing one that came to be known as "1-4-2-1."[7] That construct summarized the requirement to defend the homeland (one), engage at some unspecified but relatively low level in four critical theaters at once (four), fight and win two "overlapping" (but no longer simultaneous) major regional wars (two), and engage in one major national war if necessary (one). Amazingly enough, this requirement added up to a military of about the same size and cost as the one nominally designed to fight two MRCs in the 1990s, with a few bells and whistles added. Some might argue cynically that the shift in terminology was aimed more to conceal the fact that the military in 2001 actually could not be reasonably expected to fight two nearly simultaneous MRCs with the forces available when Bush took office (as the 2001 *Quadrennial Defense Review* admitted). Whether or not it succeeded in that aim, the 1-4-2-1 construct added further to the confusion about the military's mission that had begun to develop in 1990.

In truth, the military's mission can and should be simply stated. The United States maintains and uses its armed forces for the purpose of defending, supporting, and advancing its interests around the world. That is the basic reason all states maintain militaries. The use of the military is not restricted to situations requiring the use of force, and never has been in the United States or anywhere. Disaster relief has always been a core mission of the armed forces for the simple reason that they alone have the capability to send large numbers of people into areas with no infrastructure and maintain them there indefinitely. The Army Corps of Engineers has always existed not merely to dig ditches and build operating bases for the field forces, but also to improve America's critical infrastructure and serve civil society in a variety of ways. Thomas Jefferson established the U.S. Military Academy at West Point not only to train officers for war but also to generate a body of skilled engineers to serve the nation (as well as to ensure that a large proportion of those officers were Republicans).

The idea mooted in the 1990s that the military should "return" to its "original" purpose of "fighting and winning the nation's wars"—and, by extension, abandon the newfangled notion that it should be involved in numerous noncombatant situations—was itself the innovation, and a dangerous one. It represented an attempt begun with the Weinberger Doctrine and advanced by the Powell Doctrine to limit the tasks for which the military can be used to a set of problems the armed forces of the 1990s found congenial: primarily, major maneuver wars in which the president was willing and able to deploy overwhelming force.[8] As a result of this predilection, the armed forces spent inadequate time and effort during that decade studying the sorts of conflict in which they were regularly and properly engaged: peacekeeping operations in Europe designed to prevent the entire Balkan region from exploding; humanitarian and relief operations in Somalia (which, if pursued, could have avoided the recurring civil war there throughout the 1990s, which has led to an Ethiopian invasion of Somalia); and regular, if relatively ineffective, efforts to prevent the spread of weapons of mass destruction and to deter or punish Islamic terrorists for attacks on America and Americans.[9]

There are hints of a desire to return to the 1990s focus on wars against larger and more conventional enemies in the wake of the painful experiences in Iraq and Afghanistan. Service chiefs are increasingly hostile to sustaining the current operations in Iraq because they are straining the ground forces and drawing resources from the air and sea forces. It is easy to point out that the military would be well advised to focus on winning the war it is currently fighting rather than preparing for future contingencies that might or might not arise, but advancing the understanding of the missions for which the military should be prepared requires a more fundamental change in the nature of the discussion. It is time to stop talking about the relative priority of conventional forces, asymmetric capabilities, long-range strike, special forces, and other structural characteristics of the armed forces and time to start talking about enemies, threats, challenges, and requirements. Even the division between homeland security and the war on terror is artificial and dangerous. It reflects a desire to fence off military activities within the United States from military activities abroad, but it distorts the essential fact that the war on terror presents threats both at home and overseas that must be understood and met in a coherent way. In an era of peace, the American defense establishment became too comfortable with a national security debate that broke relatively neatly along service and departmental lines, and that trend has continued into an era of war. We must now redefine the debate to accord with reality in a dangerous world.

Priorities

Building a military requires making choices. The requirement to protect against all possible threats would bankrupt almost any state—efforts to do so did, in fact, lead to the fall of the Soviet Union. This fact has been at the forefront of American military planning since the end of World War II. Nostalgia for the supposedly "unconstrained" defense resources of the Cold War aside, military planners from 1945 on continually faced the reality that they could not afford to prepare to meet all possible or even probable threats simultaneously, and that they had to accept risk in some areas in order to reduce it in other more important or more likely scenarios. This constraint was present even during the Reagan years.

Having a clear understanding of the enemy during the Cold War made the task of prioritizing threats relatively easy, as we have seen. Maintaining an adequate nuclear deterrent was always a nonnegotiable basis of American military planning. Maintaining a conventional force strong enough when joined with that nuclear deterrent to protect Western Europe and Japan was also nonnegotiable. There was not a long-lasting consensus on much else. For long stretches, the United States accepted considerable risk in the Middle East and in the Pacific, and even, for a time, in Central and South America. But the basic and almost universally accepted requirements of deterrence and of defending our major allies gave a fairly obvious and natural shape to the U.S. military for half a century.[10]

All that changed with the end of the Cold War. The fall of the Soviet Union and the near-simultaneous Gulf War created a situation in which most American strategists saw no current enemies.

Discussions raged about which potential threats should receive priority when it came to allocating resources, but almost everyone agreed that the United States faced no active enemies against which it must be prepared to fight and win.[11]

As it happens, that view was mistaken. Al Qaeda was formed in the wake of the Soviet withdrawal from Afghanistan, and Osama bin Laden came rapidly to the conclusion that he must take his fight to the American homeland. He demonstrated his determination to do so with the 1993 bombing of the World Trade Center, which did not receive the attention it should have primarily because it was unsuccessful. Subsequent attacks on the American embassies in Africa in 1998 and the USS *Cole* in 2000 demonstrated a shift in the pattern of al Qaeda's activities that was noticed only by a few. It did not, however, reflect any fundamental shift in bin Laden's determination to strike America directly, as is now clear. At all events, the United States did face a determined and skillful enemy in the 1990s, but neither Democratic nor Republican administrations recognized that fact, largely because the existence of an organization actually at war with the United States did not cohere with the notion of a strategic pause, and was therefore dismissed.

The situation now should be clear to all. Al Qaeda continues to be at war with the United States, whether or not we choose to be at war with al Qaeda. Its leaders constantly reiterate their determination to destroy our allies throughout the world and then to destroy us. They continually reaffirm the centrality of the war in Iraq to their struggle, and their desire to use Iraq as a base for further expansion and violence. They constantly back their words with actions throughout the world, including attempts—fortunately unsuccessful to date—to attack the American homeland. It is clear that al Qaeda continues to work actively to acquire weapons of mass destruction in some form, and that it will not hesitate to use them if it does acquire them. The chlorine bombs used for a time by al Qaeda in Iraq as well as material found in computers seized in Afghanistan and elsewhere give evidence of this almost maniacal determination. Al Qaeda and its associated organizations are not a "threat"—they are an active enemy engaged in hostilities against the United States and its allies.[12]

The difference between enemies and threats is profound. Enemies are those who have declared themselves openly against us and are actively fighting us and our allies. Threats are states or organizations that are hostile to us, possibly even preparing to fight us, but not yet openly in conflict with us. Enemies require immediate and continuous response, because they are engaged in fighting us now. Threats represent potential but uncertain danger. Imminent threats require precautions, preparations, and possibly even preemption. But they remain contingent dangers, whereas enemies represent actual, certain, and present dangers.

The distinction does not necessarily mean that enemies should always receive priority over threats. Al Qaeda is a dangerous enemy, to be sure, but a full-scale war with China, for instance, would be far more dangerous and costly to America and its allies. The comparative damage that threats and enemies could do to the United States and its friends, as well as the likelihood that threats will become active enemies, must be factored into the equation when allocating resources among various priorities.

The key point, however, is that the U.S. military was redesigned in the 1990s to face numerous possible threats at a time when it was thought that we had no enemies.[13] As we consider the future of American land power, we must begin from the understanding that we face five active enemies in addition to an expanding array of threats. The many discussions of defense reform that attempt to skip over the wars in Iraq and Afghanistan and design armed forces for the "postwar" period are worse than useless. They perpetuate a peacetime, strategic pause mind-set that will lead America into great danger in this time of active war.

Enemies

America today faces five active enemies who are putting combatants and resources into the field against our military and/or attempting to launch terrorist attacks against us and our allies: al Qaeda and its associated movements in Iraq, Afghanistan, and around the world; the Jaysh al Mahdi (JAM—Moqtada al Sadr's Shi'a militia) in Iraq;

non–al Qaeda Sunni Arab insurgents in Iraq; Iran (which is providing weapons and advisors to JAM, other elements in Iraq, the Taliban in Afghanistan, and al Qaeda); and the Taliban in Afghanistan and Pakistan. These groups have declared war on the United States in one way or another, and are killing our soldiers and being killed by them on a regular basis. We cannot responsibly consider American national security strategy without starting by addressing the enemies fighting us today.

The non–al Qaeda Sunni insurgents in Baghdad are local foes, remnants of a regime deposed by American arms, who are resisting the establishment of new political order in Iraq. The Sunni Arab insurgents in Iraq receive minimal support from outside Iraq, apart from the global assistance given to their erstwhile al Qaeda allies. This group poses a fairly traditional counterinsurgency challenge to the United States and its allies in Baghdad, and it would be neither particularly difficult to defeat nor dangerous beyond Iraq's borders without the presence of international terrorist and state support for its activities.

The Taliban is supported by Pakistan, as it always was, as part of Islamabad's efforts to create and maintain its strategic depth by ensuring its influence in Afghanistan, but it has become nearly indistinguishable from al Qaeda in Afghanistan and even in Pakistan (an enemy we will consider below).

The Jaysh al Mahdi is a more complex and challenging foe. Internally riven and, in fact, engaged in fratricidal struggles, JAM is no longer entirely responsive to its nominal head, Moqtada al Sadr. It is increasingly the force of choice for Iranian intervention in Iraq, which takes the form of supplying it with weapons and advisors and, possibly, working with Sadr to bring this wayward fighting force more reliably under his control.[14] The strength of the Sadrist movement would make it more challenging to defeat than the Sunni Arab insurgents even without this external support, but its internal struggles and partial participation in the Iraqi political process offer more levers to use against it than the United States has been able to find until very recently against the other two insurgent groups.

Iran poses a challenge of an entirely different order. Iran is a major regional power attempting to expand its influence within the

Middle East. Its population is large and its oil wealth is considerable, but its efforts to translate these assets into significant military power or economic predominance have been relatively ineffective. From a conventional military standpoint, Iran has virtually no ability to project power beyond its borders, and would be easily defeated by American military power in a direct confrontation. Teheran has therefore pursued a more sophisticated approach to acquiring and maintaining influence. It has become the archetype of the state-based asymmetric threat, relying on a portfolio of sophisticated and hierarchical international terrorist organizations to advance its goals and working actively to acquire nuclear weapons technology even as it supports armed proxies fighting U.S. soldiers in Iraq with weapons and advisors.

Its support to armed groups in Iraq ranges Iran among the list of America's enemies—it is not merely a threat because its military personnel are directly engaged in an effort to defeat the United States and its allies in an ongoing war. Most of the Iranian challenge remains *in potente*, however. Iranian interference in Iraq is significant, but it is not the primary cause of instability there. Iran does not have reliable control over any of its proxies in Iraq, and has only recently appeared even to have chosen one preferred proxy— Moqtada al Sadr—over other rivals it has been supporting (and continues to support). Hostilities against Iran, therefore, fall into fairly traditional categories of Cold War–style escalation scenarios. American forces have the options of responding symmetrically to Iranian challenges in Iraq by fighting Iran's agents and proxies there, or responding asymmetrically by attacking Iran directly.

The existence of active hostilities between Iran and the United States and its allies, albeit at low levels, greatly complicates American efforts to deal with the longer-term threats the Iranian regime poses to U.S. interests in the Middle East. Iran's proxy war against the United States in Iraq means that efforts to curtail the Iranian nuclear program, to reduce Teheran's support for Hizbollah, or to limit Iranian interference in Afghanistan must all be weighed against the dangers of escalation in Iraq. At the same time, ongoing hostilities in Iraq offer potential leverage against Teheran in other areas, if it became clear that American efforts to stabilize the situation in Iraq were likely to succeed. The key point is that Iran's status as an active enemy makes the challenge of dealing with Teheran's efforts at nuclear proliferation or its support for terrorism different from almost any other scenario America has faced in places like North Korea or Libya, with which hostilities have not been going on actively for a long time. Since America's conflicts with Iran are by no means limited to the theater in which hostilities are occurring, we will consider the nature of the Iranian challenge in more detail below, in the section on threats.

By far the most dangerous and important enemy currently fighting the United States is al Qaeda, of course. Al Qaeda is a global enemy that has avowed its intention to destroy the United States and all of its allies around the world. Al Qaeda pursues an ideology that it has clearly articulated, and that is not just words. In almost every place where al Qaeda cells have achieved a degree of local control, they have begun to implement their intolerant and extreme distortions of Islam. Al Qaeda cells have established *sharia* courts and punished violators of their precepts in Afghanistan and in multiple locations in Iraq, including Anbar and Diyala provinces, Sunni Arab neighborhoods in Baghdad such as Ameriyah, and the southern belt around Baghdad, particularly Mahmudiyah, Yusufiyah, and the other small towns in that vicinity. As with Soviet Communism, it would be unwise to discount the statements of intent that al Qaeda and its affiliates regularly make, in view of the organization's repeated and determined efforts to follow through on those statements.

Al Qaeda is an unusual enemy in some respects, but a very familiar one in others. It is a revolutionary organization based on Leninist principles of organization and activity. Its name, Arabic for "the base," reflects its image of itself as the vanguard of the Islamist revolution, just as the Bolsheviks described themselves as the vanguard of the proletariat. The ideological writings of Sayyid Qutb, the intellectual forefather of al Qaeda, are heavily laden with Marxist-Leninist concepts adapted for a fundamentalist Islamic religious ideology. The repeated statements of Ayman al Zawahiri, Abu Musaab al Zarqawi, and his successors in Iraq, such as Abu Ayyub al Masri and others, all declare the organization's intention to take over a Muslim state, preferably in the heart of the Middle East, and use it to establish a base for subsequent conquest. They refer to the reestablishment of the Caliphate, by which they mean the reunification of all Muslims under a single temporal ruler (who rules strictly in accord with Muslim law and tradition as they interpret it), but their aim is actually broader than that. When they speak of the Muslim world, al Qaeda leaders do not confine themselves to the traditional boundaries of the Caliphate or even the Ottoman Empire. They include all Muslims everywhere—even the large Muslim populations of Europe and the United States. Nor do they mean to leave non-Muslims alone. They frequently refer to destroying the United States, whose culture they see as inherently destructive of Muslim religion and virtue, and to requiring non-Muslim peoples to accept either Islam or the submissive role of infidels obedient to Muslim rulers. These goals were first laid out as a coherent action program by Qutb, and they have been repeated by his followers in many lands. There is no reason to believe that al Qaeda is not sincere in its desire and intention to achieve them.[15]

In this regard, the struggle against al Qaeda resembles the struggle against Communism. Al Qaeda is a revolutionary movement within the Muslim world seeking to gain control of one or more Muslim states either by force or by persuasion or both. It works aggressively to increase the popularity and force of its message through writing and public presentations as well as through the success of its arms. The good news is that almost everywhere al Qaeda has managed to implement portions of it socio-religious program, local Muslims have resisted. The Taliban and al Qaeda often had to resort to force to sustain their policies in Afghanistan, as have al Qaeda cells in Iraq. In both countries, the popularity and local support for al Qaeda have tended to decrease over time as the socio-religious program is implemented more thoroughly. These trends strongly suggest that there is relatively little enthusiasm within the Muslim world for the actual socio-religious aims of al Qaeda, and it would be surprising to see an outpouring of support for that movement on the basis of its religious rhetoric and program.[16]

This fact suggests that the United States should be very wary of identifying the current fight as a struggle against Islam or the Muslim world. President Bush has been very careful to make clear that the fight is against a particular radical revolutionary ideology within that world, while avoiding the temptation to reach for more simplistic metaphors like the "clash of civilizations." It is worth noting that the Bolshevik ideology was tremendously unpopular in Russia prior to the October Revolution—and even long after that. The Bolsheviks did not become a mass party until after the collapse not only of the Tsarist regime, but also of the much more democratic Provisional Government that followed. If Al Qaeda's leaders understand the degree of disdain in which their religious views are held by most Muslims, they do not care. They are following the Leninist playbook very closely, concentrating on seizing power in a weakened Muslim state and imposing their ideology by force upon the people. They then intend to use that hapless land as a base for further conquest, just as the Bolsheviks did once they attained power in Russia. We have the tragic history of that period as evidence that success in such endeavors is possible.

All of this means that the analogy between the current conflict and the Cold War is problematic. The current period is much more similar to the struggle against Communism prior to 1917, when the Leninist movement was very small, entirely revolutionary, organized into cells, and linked with other groups internationally primarily by its ideology. It is not like the period after 1945, when Communism had become almost identical with the Soviet Union, a large, cautious state whose rulers prioritized national survival over ideological advances. Deterring the Soviets was feasible; deterring Lenin was not. In the 1980s, the United States found out how to defeat an ideological movement that had already become identified with a powerful state. We now must figure out how to defeat an ideological movement before it takes control of a state—a task that is much more difficult.

One other important difference between the current struggle and the Cold War is that this war is hot. Al Qaeda is attacking Americans every day in Iraq and Afghanistan, has already attacked the homelands of the United States and its allies in Spain, Britain, and Jordan, and has been prevented from conducting additional attacks through vigilance and good fortune. It is self-evidently impossible to deter this organization from attacks that it is conducting and planning to conduct regularly. We will have to fight it.

We will consider in the next chapter what sort of military operations are best suited to fighting al Qaeda, but here it is essential to note one thing. Just like Bolshevism, al Qaeda is a disease that feeds on weakness. Collapsed states in the Muslim world offer opportunities that al Qaeda cells seek out in an effort to gain a foothold and, ultimately, control. The international community allowed this process to move to completion in Afghanistan in the 1990s, and paid a horrible price. Al Qaeda cells have targeted weak or collapsed regimes in Somalia, Lebanon, and Algeria. The process was underway in Iraq until very recently, when al Qaeda mistakes and skillful U.S. strategy began to turn the tide. But allowing Iraq to fall back into total collapse will restore al Qaeda's prospects for success there, hope of which is fueling the continued al Qaeda surge against American efforts.

Because al Qaeda is concentrating so many of its global resources on Iraq, defeating it there is clearly an essential precondition for any kind of success in the current struggle. Other al Qaeda cells appear to be bidding for prominence within the movement, however, and we must watch these developments very carefully. Al Qaeda in the Islamic Maghreb has recently conducted several attacks in Algeria, although these have yielded relatively little benefit to those cells so far. Al Qaeda cells in the Palestinian camps in Lebanon provoked—probably unintentionally—a major confrontation with the Lebanese government that they appear to have lost for now. These are the more visible dangers at the moment, but others will surface over time. Identifying them, recognizing their potential danger and power, and determining the best ways to defeat them will be an essential component of any successful American strategy in the current war and a core mission for the U.S. armed forces. But let us be clear: failed and failing states in the Muslim world are powerful attractors of al Qaeda attention. We probably do not need to address every one, but we do need to be able to intervene anywhere we see al Qaeda cells establishing deep roots in the carcass of a decaying state.

Threats

Beyond the enemies of the day, there are a certain number of states that pose significant threats to the United States and its interests by selecting objectives antithetical to ours and developing the military and economic means to pursue those objectives. Of these states Iran presents the most imminent challenge. North Korea is rapidly fading as a threat, although it continues to be a significant challenge. China is potentially the greatest likely threat, although the probability of direct conflict with China in the near term is very low. Venezuela has the hostile intention and the economic resources to be a significant regional threat, but its military forces remain rudimentary, and its ability to attack U.S. interests directly is very low. Russia has been playing an extremely negative role in the world, and the statements of President Vladimir Putin suggest that this trend is likely to continue, but overt conflict between the United States and Russia in the near- to mid-term is extremely unlikely. Internal challenges in Pakistan, Saudi Arabia, Indonesia, and other Muslim states are important and will be considered in the next section, but none of these states poses a direct threat to American interests at the moment. By far the most immediate and potent threat, therefore, is Iran, with China a distant second only because of the unlikelihood of direct conflict in the near future.

Iran. Successive Iranian governments since 1979 have proclaimed objectives that are antithetical to core American interests in the Middle East and around the world. Iran's nuclear program began under the presidency of Ayatollah Akbar Hashemi Rafsanjani—now touted as a reformer that some in the United States expect to be more "reasonable" about the pursuit of this program. Iran's anti-Israel rhetoric has been consistent, even if Mahmoud Ahmadinejad and his colleagues have increased its prominence in recent years.

Above all, Iran's aim to establish itself as the hegemon of the Middle East is clear, well developed, and completely unacceptable to the United States.

The Iranian leadership recognizes that it cannot defeat the United States in a conventional struggle. It has therefore adopted a multipronged approach aimed at protecting itself from American military action, driving the United States out of the Muslim world, and making itself the arbiter of the lands between Pakistan and the Mediterranean at least. It seems clear that Teheran's pursuit of nuclear weapons is designed to eliminate the danger of an American military strike.

The twin lessons of Iraq and North Korea have sent a powerful signal to Iran and other rogue states: any state that is seriously at odds with America is in grave danger without a nuclear arsenal; with one it is not only safe, but in a strong bargaining position. Saddam Hussein was known not to have nuclear weapons in the 1990s, but wished to continue policies that put him at odds with the United States. His weakness and obstinacy led to repeated military strikes and, finally, the invasion of 2003 that removed his regime from power. North Korea's conventional military threat to the south was enough to preserve that regime from attack in the same period, and its partially successful test of nuclear weapons has been enough not merely to take the threat of U.S. military action off the table, but to induce the United States into ever higher levels of diplomatic engagement of the sort Pyongyang has been seeking for years. Iran's leaders seem to have absorbed these lessons well, which may explain their determined pursuit of nuclear weapons as the United States appears weak and embroiled in conflicts on their borders. There is no way to tell if the periodic threats Iranian leaders growl at Israel should be taken as a sign that they would use nuclear weapons against the Israelis as soon as they had them, but it is more likely that they intend primarily to procure a de facto guarantee against American military action as they expand their hegemonic ambitions in their region.

Iran's main focus, beyond the nuclear program, appears to be in Iraq. Iraq, after all, is the only state in recent history that has posed an existential threat to Iran (during the Iran-Iraq War, which is as seminal an event in modern Iranian history as World War I was in Europe). Iran's behavior in Iraq demonstrates its objectives clearly. Teheran aims first and foremost to drive the United States out of Iraq in defeat. Secondarily, it seeks to keep Iraq weak. Finally, it is looking for reliable proxies that will ensure that Iraq is at least friendly toward, if not allied with, Teheran.

Iranian agents have been providing funds and weapons profligately to almost every group in Iraq fighting against the United States and the current Iraqi government. Iranian support flows to all Shi'a groups, but also to al Qaeda and other Sunni Arab groups that pursue objectives at odds with Iran's goals.[17] Al Qaeda, especially as it has developed in Iraq, is a powerfully anti-Shi'a movement; the Islamic State of Iraq (al Qaeda's front group) repeatedly condemns not merely the United States, but its Shi'a allies, whom it regards as dangerous heretics and schismatics. But Iranian support has been flowing to this group nevertheless, and the reasons can only be that it is a powerful enemy of the United States and perpetuates disorder and weakness in Iraq.

After all, if Iran simply wanted a stable Shi'a state in Iraq, it would be working actively to support the current government there: Nuri Kamal al Maliki is the head of the Dawa Party, a Shi'a organization with no history of hostility to Iran, and his power rests on the Supreme Islamic Iraq Council (formerly the Supreme Council for the Islamic Revolution in Iraq), a group formed of Iraqi exiles in Iran during the Iran-Iraq War. If Iran sought only a friendly, stable Shi'a government in Baghdad, it would hardly be funneling weapons to al Qaeda fighters who are killing Shi'a on a daily basis, nor would it be building up the Jaysh al Mahdi, the least reliable and most dangerous Shi'a military force in the country. The Iranian aim clearly goes beyond ensuring a friendly Shi'a government in Baghdad. Teheran seeks to defeat and humiliate the United States. in Iraq for its own purposes, and to ensure that it has either a reliable proxy in power there or chaos that cannot threaten the Iranian state.

Iran will therefore remain a potent threat even to a successful Iraq. Unless the Iranian leadership decides to live with an independent, democratic, stable Shi'a state that is not under its control—

something it has shown no sign of desiring or being willing to tolerate so far—Iraq will face a significant challenge to its east for years to come. The United States will therefore have to play a critical role in protecting its fledgling ally long after the insurgency in Iraq has been brought under control and the government is functioning once again.

Iran's aims go far beyond Iraq. Teheran has been a major state sponsor of terrorism since the 1979 revolution, and it is expanding its portfolio of terrorist clients. Its Hizbollah proxy in Lebanon recently succeeded in humiliating the Israelis, although it has failed to deter the Lebanese government from supporting international demands for a tribunal that is nearly certain to find Syrian complicity in the assassination of former prime minister Rafik Hariri. The recent triumph of Hamas over Fatah in Gaza is another victory for Iran, which has taken Hamas under its wing, much to the consternation of many Arab states. The growth in power and prestige of Iranian-supported terrorist movements in the Levant (where Nasrallah has become a popular name for children) is a dangerous trend. The movement of Bashar al Assad's Allawite government in Syria closer to Teheran is another. Iran's increasing involvement in Afghanistan—again, working to destabilize a democratically elected government supported by the United States—shows that its aims are not restricted to the Middle East.[18]

Full-scale war with Iran is not inevitable or even advisable at the moment. But the trend lines are ominous. There is no sign that Teheran is seeking any sort of accommodation with the United States. On the contrary, it is increasing its aid to America's enemies in Iraq and Afghanistan and to terrorist organizations far beyond its borders. Attempts to deter Teheran with military force and economic sanctions do not appear to have been successful, considering this steady growth in Iranian efforts throughout the region. These attempts have been undermined dramatically, of course, by moral and practical support given to Teheran by Moscow and Beijing. The United States must recognize that conflict with Iran may become necessary, or may develop organically out of any of the several issues already in conflict. The military must be prepared to accept the mission of defeating Iran in a responsible manner that does not create uncontrolled chaos in the region—something that it is not now capable of doing.

China. The threat from China is potentially very large, but also complex and unclear. Fortunately for this study, it does not weigh very heavily on American land power capabilities. It is quite possible to imagine scenarios in which American soldiers and marines had to help defend Taiwan from a successful Chinese amphibious invasion, but the preferred solutions to that problem remain deterring the attack in the first place and thwarting it before Chinese forces get to ground—both tasks that are almost exclusively the province of the navy and the air force. Like Iran, China poses a more complex threat to U.S. interests in its region, parts of which could conceivably involve significant numbers of American ground forces. China's predominant influence with North Korea could produce very interesting scenarios if the current regime in Pyongyang were to collapse. It is also possible to imagine more far-fetched scenarios in which U.S. and Chinese forces engage either directly or by proxy in Central Asia, where both powers have increasing interests.

But the main impact of the Chinese threat on American land power lies in the fact that it forces the United States to choose between developing the large numbers of sophisticated air and sea weapons needed to fight China in any likely conflict, and building the large ground forces needed in ongoing conflicts and the most likely wars of the near future. In truth, there is no way to make such a choice responsibly. American interests require the defense of Taiwan, Japan, and South Korea, and so require the United States to be able to deter China from adventurism and to defeat Chinese aggression should deterrence fail. At the same time, the United States must be able to support the wars it is currently engaged in and to prepare for possible future conflicts that might come whether or not we would prefer to fight them.

The bottom line for the United States is the need to recognize that the world is fundamentally different from the mirage of the 1990s that so clouded our thinking on defense. We cannot ignore the wars we are actually fighting in order to prepare for possible future conflicts, nor can we fail to prepare for those conflicts in order to focus exclusively on our ongoing struggles. America is the world's only remaining global power. In truth, it has been the only real global power since the collapse of the British Empire. The Soviet Union never really had either the capability to project power, or the interests to defend and pursue around the world, that America has. No state in the world today even comes close.

Recognizing the true nature of the world today means rejecting the trade-offs demanded of the military in the 1990s. Current conflict and preparation for future war cannot constitute a zero-sum game. The holiday of the 1990s is over, and the coming decades look dangerous rather than hopeful. The U.S. military must be able to execute ongoing missions, be prepared for likely future missions, and be ready for remote contingencies that require long lead times for our preparation. All of that means that the defense budget must rise dramatically and for the foreseeable future to encompass all of these nonnegotiable requirements. The calls for a significant increase in ground forces laid out below, therefore, must not be understood as calls for reductions in the support of America's naval and air forces. If America wishes to defend its vital interests around the world now and in the future, it must be willing to pay to do so.

Challenges

In addition to enemies and threats, the United States faces a number of current and potential challenges for which it must be ready. State collapse in North Korea, Pakistan, or Saudi Arabia could lead to major regional challenges and, in the first two cases, great concern about the danger of weapons of mass destruction falling into the wrong hands. The United States could not simply look on with unconcern in any of these cases, although it might well be able to avoid large-scale intervention in many scenarios. The American armed forces must nevertheless be prepared to meet these challenges should they arise.

The extremely precarious situation in the Horn of Africa poses another series of potential challenges. The United States withdrawal from Somalia in the early 1990s led to a civil war that opened the region up to further al Qaeda penetration. The Ethiopian invasion at the end of 2006 reduced that problem temporarily, but Ethiopia is incapable of maintaining order in Somalia by itself, and international bodies have been extremely slow to shoulder the burden. There is a great danger that the situation in the Horn of Africa will collapse once more, and the United States will again be faced with the choice of intervening or allowing al Qaeda to establish a base in a chaotic region.

Troubles in Indonesia, where al Qaeda has a strong and growing presence, will also require America's attention and at least low-level intervention. The United States must also keep sight of the developing situation in Latin America, where Hugo Chavez is playing an extremely negative role and the dangers of narco-trafficking have increased as terrorist groups increasingly reach out to such illicit sources of income to support their own activities. Any of several scenarios in Central America, particularly in Cuba, could require complex humanitarian interventions for which the U.S. military should be ready.

In the 1990s, many defense analysts had difficulty finding likely near-term scenarios for the use of American military force. Now we face the opposite situation. Even if America adopts an entirely reactive posture in the world, abjures preemption or preventive war, and decides to act only when acted upon, the visible and likely threats and challenges today are more than enough to strain our military capabilities, particularly our ground forces. For this discussion does not presuppose that the current administration or its successor will seek out conflict. It flows from the fact that America faces a given set of enemies today, as well as serious threats whose behavior might draw us into conflict we would rather avoid, and challenges that could easily arise and involve us despite our desire to remain uninvolved. Iranian efforts at regional hegemony, the collapse of nuclear-armed states, humanitarian disasters that generate refugee flows into the United States, situations of chaos that provide bases to our deadliest enemy—these are not challenges that the United States can simply afford to ignore. Prudent military planning requires preparing to meet them now, before they arise, so that we have the option of engaging earlier in crises and shaping them to our advantage rather than allowing them to spin out of control before we become engulfed in them.

Requirements

Leaving aside the need to respond to enemies, threats, and challenges, U.S. ground forces must perform a number of core functions both to ensure their own institutional well-being and to support the nation in ways that go beyond the use of force. Excessive cuts in the 1990s, unwise searches for efficiency at a time of national crisis, and the failure to grow the ground forces despite their involvement in two significant land wars since 2003 have hollowed out the institutional base of the army. Soldiers and officers have been transferred from training and educational institutions into line units to make up for deficiencies in combat personnel. As a short-term expedient, this shift made sense, but it cannot be sustained for very long. The army and marines must grow rapidly, even more rapidly than current proposals call for, and that can occur only through an increase in the training base in both services. In addition, the intellectual and educational task of assimilating lessons from ongoing struggles to avoid losing them—a task at which the military failed after Vietnam—requires rebuilding and sustaining the ground forces' intellectual capital at service academies, schools, war colleges, and other such programs.

The need for infantry in the current conflict has driven the search for combat soldiers even further, however. Artillery and armored units in Iraq today are trained and used primarily as infantry. Reports indicate that some artillerymen and tankers have never qualified their systems. Virtually no units in the active army or marines have trained for high-end conventional warfare in years. When the trade-off is providing the combat power necessary to win the current war or preparing for future contingencies, the choice is obvious and easy. But protracting such a stop-gap approach requires accepting considerable risk for the future. The ground forces must be able to retain their core competencies as well as support ongoing struggles. We cannot imagine that all future wars will look like the ones we are fighting now, any more than we can convince ourselves that we will never fight another counterinsurgency again. The U.S. military must remain a full-spectrum force able to move from peacekeeping to major war on short notice.

The military must also be able to meet unexpected requirements for homeland security and disaster relief. Traditionally, the army has relied upon National Guard units for such tasks, but a number of factors are making that reliance more difficult. The army has deployed many Guard units in the Balkans, Afghanistan, and Iraq, sometimes pulling them out of the areas in which disasters occur, as happened during the Hurricane Katrina recovery efforts. The increasing willingness of the senior military leadership to see Guard units as operational rather than strategic reserves will also have implications for their training and their availability for traditional missions. The trade-offs are not clear-cut. As long as America faces an urgent need for combat power in ongoing wars, then the priority must go to filling that need rather than keeping force back for possible contingencies. But the military remains the only force that can reliably and quickly respond to disasters, whether man-made or natural, on a large scale. This capability must be adequately resourced not simply as a lesser included requirement, but as a core mission.[19]

The military's mission is easy to state, but difficult to execute in the world as it is today. The U.S. armed forces—and the defense debate in general—have not adequately made the transition from a strategic pause mind-set to the wartime environment we now face. Worse still, American military budgeting tradition is working badly against doing the right thing today. Historically, the military budget falls after every major war. The military services even today are preparing budgets on the assumption that this pattern will be repeated as the Iraq War winds down one way or the other. Nothing could be more dangerous than this assumption. The mission-set facing the military today is greater than it has been since the mid-1980s and even more complicated to understand than it was then. None of these missions is optional; each involves serious risk that

places the welfare of the nation in great jeopardy. The Bush administration suffered from the challenge of having to move rapidly from
strategic pause to confusing war. The next presidential administration will enter in the midst of ongoing war, whatever happens in
Iraq over the coming year. It is time to rebuild our military thinking to accord with this reality.

2

What Kind of War?

The American armed forces have had to consider what sorts of military conflict to prepare for since early in the Cold War. Within a few years of the end of World War II, it was apparent that the Soviet Union and its allies posed military challenges ranging from nuclear attack to low-level Communist insurgencies in many areas of the world at once. Major conventional war in central Europe was a constant among these threats, and formed the basis for the organization, doctrine, and equipping of the American ground forces for most of the Cold War, but other points on the "spectrum of conflict" (as it is now called) also shaped those forces. Dwight D. Eisenhower oversaw the abortive "Pentomic Division" designed to allow ground forces to function effectively on a nuclear battlefield, and in the 1960s John F. Kennedy formed the Special Forces to address the threat of Soviet-backed insurgencies around the world. Through most of the Cold War, the U.S. military accepted a requirement to be able to fight across the spectrum, although it never had the capability to do so adequately at every level at the same time.[1]

The end of the Cold War brought no real change in this constant tension. The requirement to combat Communist insurgencies became the requirement to conduct peacekeeping operations and nation building in failed or failing states. The 9/11 attacks added a new focus on counterterrorism operations (although most U.S. military activities in the 1980s were focused on counterterrorism one way or another), but it did not eliminate the requirement to prepare for large-scale conventional or even nuclear war. Defense Department reviews in 2001 and 2005 reiterated the need for the

U.S. armed forces to be able to respond "across the spectrum" of violence from the highest end to the lowest.[2]

The trouble is that preparing for threats across the spectrum is extremely expensive, and post–Cold War administrations and Congresses have resisted proposals to increase the baseline "peace dividend" defense budget materially, apart from the costs of individual military operations and wars (now funded through annual supplemental requests rather than a normal defense budgeting process that plans for supporting continued military operations over the medium term). And the search for new paradigms by which to understand the seemingly novel challenges of the post-9/11 period has led some to advocate abandoning parts of the spectrum and focusing on the "future of war," usually described as "asymmetric" conflict, by which is meant terrorism and insurgency.[3]

At root, most such proposals are fiscal in origin: they attempt to offer ways to allow the United States to fight complex networked foes without having to maintain expensive high-end conventional forces. In some cases, these proposals originate with the idea that high-end conventional forces are unwilling and unable to do what it takes to fight complex networked enemies, and that being prepared for conventional war is antithetical to fighting terrorists. In all cases, these proposals pose a choice: the United States can maintain the capability to fight large conventional wars or it can develop the capacity to fight low-intensity, asymmetrical conflicts against networked and elusive enemies. The question is: Can or should America choose to dominate only part of the spectrum of violence? Must the United States really pay to maintain the capacity to fight all kinds of wars? Is it not possible to accept risk only in some areas in order to create an affordable military tailored to the sorts of conflict most likely to dominate the coming years?

The Nature of Conflict and Attempts to Predict the Nature of Future War

The argument that the United States should be prepared for all kinds of conflict is easy to make. America faces potential nuclear threats from Russia and China, and it is not hard to see such threats emerging in Iran and North Korea within a number of years. War against Iran, or against China over Taiwan, would be a taxing, conventional conflict, at least at the beginning. Ongoing wars in Iraq and Afghanistan combine counterinsurgency with counterterrorism. Shaping operations in the Horn of Africa, Indonesia, and elsewhere take the form of regular peacetime engagement. The trouble is that the U.S. armed forces are not now able comfortably to contemplate conventional war against Iran or even China (despite the airpower-heavy focus of any likely conflict over Taiwan) while also fighting counterinsurgencies in Iraq and Afghanistan and conducting global counterterrorism efforts. In fact, as we shall see below, the current American military posture contains an inordinate and unacceptable amount of risk all across the spectrum of conflict.

But fixing this problem would be expensive, and some say impossible, within the current organization and mind-set of the U.S. military. The armed forces, according to this view, should focus instead on the likeliest types of future conflict, usually low-intensity, asymmetrical struggles against networked terrorist groups;[4] or they should refocus on high-end threats like China.[5] These arguments are difficult to make persuasively, however. They rely on the assumption that it is possible for a state like the United States either to predict the most likely nature of future conflicts or to choose to fight only certain kinds of wars. There is considerable historical evidence against this assumption.

Western military strategists have been attempting to predict the nature of the next war for hundreds of years. (Chinese authors— think of Sun Tsu—have been at it a lot longer.) The record of such attempts is discouraging. For every Napoleon, von Seeckt, Tukhachevsky, Mitchell, or Trenchard who got it (mostly) right, history is strewn with scores of contemporaries who got it completely wrong. Sometimes every major power gets it wrong, as happened before World War I, when everyone expected a short war of maneuver and got a long, grinding war of horrible attrition. Sometimes visionaries develop the right general idea, like De Gaulle and Fuller in the interwar years, only to be shunted aside and ignored by less imaginative leaders. Sometimes brilliant insights and theories turn out to be irrelevant, like the Soviet concept of the "revolution in military affairs," created in the late 1950s to describe in detail the nature of a future nuclear war that was never fought.

More recent American visionaries have described future worlds in which long-range firepower directed by satellite intelligence would dominate all conflict—a vision that lies now in ruins on the battlefields of Iraq and Afghanistan. In the 1990s, others confidently predicted that Desert Storm was the last major armored war, that all future conflict would be nonlinear and unconventional.[6] Operation Iraqi Freedom was a well-fought but pretty conventional mechanized war. Current visionaries who claim to know what future conflicts will look like may prove right, but the historical evidence suggests that relying on the accuracy of their visions would be extremely dangerous. The project of predicting the nature of future war is simply not a reasonable undertaking beyond the realm of theoretical discussion. It cannot and must not be the primary basis for designing armed forces.

Nor can a global power like the United States choose its wars all the time. If China attacked Taiwan, American vital interests would require the United States to intervene. There are a variety of steps Iran could take that would similarly make war with Teheran essential to preserving American interests. Many scenarios involving the collapse of Pakistan, North Korea, or Saudi Arabia would jeopardize American security and/or invoke alliance commitments in fundamental ways that would require American military involvement. And, of course, wars in Iraq and Afghanistan continue, as do counterterrorism and peacetime engagement operations around the world.

Although war with China or North Korea, intervention in Pakistan or Saudi Arabia, or even conventional conflict with Iran seems moderately to very unlikely, responsible strategists have to acknowledge that the United States might need to engage in such conflicts in the future. There was probably never a time in American history before September 11, 2001, when anyone thought that the United States might fight a significant war and then a protracted counterinsurgency/peacekeeping operation in Afghanistan—tellingly, the United States Central Command (CENTCOM) had no war plan on its shelves for any such contingency on 9/11, and one had to be developed from scratch.[7] Almost no one could contemplate in 1910 the likelihood that America would send hundreds of thousands of troops to fight in France within eight years. Even in January 1950 the idea that within two years the United States would fight not only North Korea but China as well seemed absurd. It is simply not possible to rule out certain kinds of conflicts in advance, no matter how unlikely they may seem at any given moment.

The costs of American unpreparedness to fight each of these wars (and most other wars as well, for that matter) were staggering. The scramble to create an army from scratch in 1918 (and nearly from scratch in 1941) led to tens of thousands of unnecessary casualties, not to mention enormous setbacks in the first months of the struggle. Failure to maintain adequate military power after World War II led to the humiliation of Task Force Smith in Korea in 1950—and the very near collapse of the entire American effort in the peninsula within weeks of the North Korean invasion.[8] The consequences of failures of preparation and planning for the wars in Afghanistan and Iraq in 2001 and 2003 are known to all. The United States simply cannot afford to build a military on the assumption that it will not have to fight particular kinds of wars.

Neither can the United States be perfectly prepared to fight every kind of war with no notice, of course. Any such military policy would cost far more than Americans are or should be willing to pay. The challenge of designing the U.S. armed forces now lies in recognizing and properly balancing the cost of military preparedness against the risks assumed in choosing to be less than fully ready for foreseeable contingencies. Those risks must be measured along two axes—the likelihood of the contingency, and the consequences of being unprepared for it. Any such discussion, however, must begin with an evaluation of the risks the current American military posture currently assumes.

The Posture of the U.S. Military Today

The United States is currently engaged in two large-scale counterinsurgency/counterterrorism wars and maintains small but long-term presences in two peacekeeping operations (Kosovo and the Sinai). It is committed to a long-term deterrence/engagement mission in South Korea, and is engaged in a global counterterrorism effort against al Qaeda. The sum of these ongoing deployments weighs disproportionately on the ground forces—the army, marines, and Special Forces. Air force and navy aircraft and ships support all these efforts, but this support occupies a relatively small part of those services' capabilities at any given time compared to the burden placed on the land forces.

The ground forces are fully engaged. U.S. Special Forces assets are almost entirely committed to the wars in Afghanistan/Pakistan and Iraq, to the detriment of their operations in other theaters. At the moment, the twenty-one brigade-equivalents of combat power engaged in active combat operations in Iraq comprise half of all the brigade combat teams in the active army. The burden is spread somewhat through the use of the equivalent of three marine regiments (their equivalent of brigades), but the cost is a dramatic reduction in the marines' ability to respond rapidly to unforeseen contingencies—their core mission. A small number of National Guard brigades conduct security missions in Iraq, but most deployable guard brigades have already served at least once in Iraq and/or Afghanistan and are not seen as being deployable again in the near term.

The strain on the ground forces is greater than these basic numbers suggest, moreover. The United States maintains more than twenty thousand troops in Afghanistan, many of them high-demand/low-density combat service support units, and both active duty and guard units are needed in Iraq to protect bases and convoy routes throughout a large and dangerous country. The military advisory effort in Iraq also places a surprising burden on the ground forces. Advisory teams consist of officers and senior NCOs—typically from captain to colonel and sergeants. Put another way, each team of one colonel, one or two lieutenant colonels, several majors and captains, and a few sergeants is the equivalent of the command team of a U.S. Army brigade. There are now around six thousand personnel in such teams in Iraq—the equivalent of many brigades' worth of senior leaders. It is worth noting here that shifting the U.S. presence in Iraq from a counterinsurgency to an advisory role would increase the pressure on these already overstretched ranks. The idea of expanding the advisory effort to twenty thousand experienced, skilled, and trained officers and NCOs, as some have suggested, is probably unrealistic given the actual pressures on those ranks in the force today.[9]

Because the Bush administration has refused until very recently to contemplate increasing the size of the ground forces, the army and marines have had to support these strains out of a fixed pool of resources. They have responded by deploying every combat unit in the active force, including units based in Korea and units that were serving as the permanent "opposing force" at combat training centers. The army has hollowed out its institutional base force, drawing officers and senior NCOs from its educational, training, doctrine, and other key base units to fill out combat units. The dramatic increase in the use of contractors—recent reports indicate that there are now more contractors in Iraq than soldiers—is another result of the strain on the ground forces.

Equipment shortages have created another set of constraints. Active units preparing to deploy to Iraq have often had to cannibalize other units to fill out their equipment sets. The situation in the National Guard is very dire in this regard—few National Guard units could deploy on short notice with anything like the required equipment; many could not find enough personnel to fill out their rosters.

In addition, contraction of the army's training base over the years limits the speed with which the force could be expanded now. Lack of funding has also completely destroyed the military's ability to undertake conscription, should it become necessary on short notice. There is no real mechanism now in place at any level to induct and train new soldiers, either volunteers or conscripts, in large numbers.

In short, the American ground forces face an unprecedented crisis. In the midst of two wars, neither of which would permit a responsible reduction in forces in the short term, the ground forces cannot contemplate another conflict, cannot expand rapidly, and have no significant reserves either at the operational or at the strategic level. If war came with Iran, if Pakistan or North Korea collapsed, if conflict spread in the Middle East, the United States would face horrendous choices: allow unacceptable crises to grow unchecked; or draft and send untrained, unprepared, and ill-equipped soldiers to the front; or use unconstrained airpower to inflict massive casualties on our foes without being able to follow up those attacks with ground forces to control the situation. Considering that the United States is already engaged in a proxy war with Iran in Iraq and Afghanistan, that Iran is known to be supporting al Qaeda and pursuing a nuclear weapon, and that the United States is not entirely in control of the potential escalation of that tension, the risk inherent in the current military posture is completely unacceptable.

We have argued elsewhere that large-scale and rapid withdrawal of combat power from Iraq is unwise and dangerous, and we will not revisit that argument here.[10] We must note, however, that even an immediate and complete withdrawal of U.S. forces from Iraq (something that no responsible leader is advocating) would not make the risks assumed by the current military posture acceptable. Iraq and Afghanistan are both countries with populations of around twenty-seven million people. Iran's population is over seventy million, its landmass is three times that of Iraq, and its terrain is vastly more complex and difficult. Ground forces operations in Iran (or Pakistan) would require many more soldiers than are currently deployed in Iraq—a deployment that is badly straining the ground forces. That is to say nothing about the difficulty of pulling out of Iraq in defeat, retraining and reequipping a humiliated army and Marine Corps, and then throwing them back into the fight in the Middle East or South Asia on a larger and more challenging scale. Even if it were possible responsibly to "see past Iraq" in planning the U.S. military posture—which it most emphatically is not—the risks inherent in the current posture are simply untenable unless we decide, with no basis in reality or history, that we will not have to fight certain wars.

The Need for a Full-Spectrum Force

At the beginning of the post–Cold War drawdown, Senator Sam Nunn argued that the United States could responsibly reduce its military because it would be able to "turn inside its enemies," re-arming faster than any potential foe if tensions and crises seemed to warn of impending conflict. This argument, among others, provided key support for the unwarranted decision to cut the American military by around 40 percent across the board in the wake of the Soviet Union's collapse. It has proven to be disastrously mistaken.[11]

One of the key lessons of the last seven years has been the astonishing inertia and inflexibility of the American military posture. After four years of visible strain on the ground forces, the number of active duty troops has been increased by only thirty thousand— and that on a temporary basis until this year—an increase of around 4.6 percent of the total active duty army and marine strength. After four years of deaths due to improvised explosive devices planted like landmines—and growing in sophistication— the Defense Department has only just determined to purchase large numbers of mine-resistant vehicles to replace highly vulnerable Humvees. And that decision came after repeated scandals about the delays in procuring up-armoring kits to protect basic Humvees. All the while, the army leadership in the Pentagon has focused heavily on protecting its Future Combat System (FCS). The main ground combat vehicles called for under that program—those that would replace the M1 Abrams tank and the Bradley fighting vehicle—were not designed for the war in Iraq and will not be procured and fielded for more than a decade; they will have little impact on that conflict. Some other aspects of the complex program, however, such as the systems meant to link ground forces in a network and the robotic vehicles used to clear buildings more safely, will be of immediate benefit. The other services have fought to defend advanced aircraft and ship designs of no relevance whatever for current conflicts. The evidence is overwhelming that the current U.S. Defense Department is far from agile and extremely unlikely to be able to support a rapid turn "inside of" any enemy.

Many proposals abound for improving the agility of the Defense Department, and some no doubt have merit. But the Defense Department is not the only or even the primary culprit in this inflexibility. Congress is historically resistant to expedited equipment purchases that require waving the Byzantine labyrinth of regulations it has put in place to ensure that not a single penny is wasted in defense procurement. And the Bush administration—like the two administrations before it—has been extraordinarily reluctant to abandon key domestic goals in order to finance dramatic military expansion or even change. None of these conditions is likely to go away, even if the agility of one of the largest bureaucracies in the world could somehow be increased.

All of which is to say that the American people should have no confidence whatsoever in the ability of their armed forces to expand or change rapidly in response to developing threats or growing crises. Considering how imminent and advanced some of today's threats and crises are, moreover, we are well past the point at which prudent strategists would already have begun to change their militaries in response. Donald Rumsfeld was widely criticized for declaring that you go to war with the military you have, considering his role in creating that military. On the one hand, the criticism was valid—he had not been trying to create any other sort of military. On the other hand, it was not valid—he would have been very hard-pressed to change the armed forces dramatically even if he had been trying to do the right thing to begin with. It is very likely that the United States will have to fight the next war with the military it has at the start. This is all the more reason to develop and maintain now the armed forces that will be necessary for success in a variety of possible conflicts across the spectrum.

The Internationalist Chimera

It had been commonplace in some circles until very recently to argue that the United States should not try to prepare to fight future wars alone. America has many allies around the world, and critics of the Bush administration often argue that it was only Bush's inept diplomacy that laid the burden for Iraq so heavily on America's armed forces. This argument is not new. Eisenhower used it in the 1950s to argue that the United States should focus on its nuclear arsenal and airpower and rely upon its allies to provide ground forces.[12] And

Rumsfeld famously used a version of it in Afghanistan, where the initial American involvement consisted of a small number of Special Forces calling in strikes from American air assets in support of indigenous ground forces.

We have argued elsewhere that the dangers of relying on indigenous allies in complex postconflict situations often far outweigh the advantages of minimizing the American ground force presence.[13] But we must also note here that even the best diplomacy in the world cannot overcome a fundamental fact: America's allies have disarmed themselves, on the whole, even more than the United States has. The United States is now spending around 4 percent of its GDP on defense. NATO set a target of 2 percent for its European members—and hardly any of them are meeting that goal. The two best armies in Europe—the French and the British—number fewer than two hundred thousand soldiers combined. The French reckon on being able to deploy twenty-five thousand of those to a *major* contingency, and have no reserve whatsoever.[14] The British deploy a far higher percentage of their force and have a reserve—the Territorial Army—but British deployment includes thousands of soldiers in Northern Ireland, and its Territorial Army forces are as overstretched and burned out as the American National Guard. Germany's Bundeswehr is large, in principle, but it is still a conscript force, trained to a fairly low standard, and bound by constitutional limitations and serious national caveats that make it hard to deploy and harder to use. The other NATO states that maintain deployable forces—Italy, Spain, and Poland, primarily—can send anywhere from one to three thousand soldiers abroad at a time. Such contributions are useful and important, to be sure, but they cannot in any way make up for inadequacies in the American military.

Beyond NATO, Japan has begun strengthening its military and reducing the restrictions on its use, but its attention will remain on its immediate environment for a long time to come. Even an expanded Japanese military is unlikely to send large numbers of combat troops to the Middle East or South Asia. South Korea has also contributed about a brigade to the stabilization of Iraq, and seems set to continue that contribution, but it is hard to imagine

Seoul dramatically increasing its participation in this or any future out-of-area war. Australia and Canada have made important contributions, especially in Afghanistan, but also in Iraq—but their armies are tiny and, given their populations and respective economic conditions, certain to remain so.

One might consider reaching out beyond American allies to the large Indian Army, or to Muslim states like Pakistan, Burma, Indonesia, or Egypt, for assistance in policing Muslim lands in the Middle East; but the political problems of any such undertakings would be daunting. The tenuous relationship between India and Pakistan would be badly strained by significant Indian deployments to Muslim states to the west of Pakistan, to say nothing of the strains within Indian society between Muslims and Hindus. The idea of deploying the unstable Pakistan Army, at once central to maintaining secular government in Islamabad and a danger to that goal because of its growing infiltration by Muslim extremists, is clearly unwise. Indonesia faces its own domestic challenges with Muslim extremists, as does Egypt. One should consider long and hard the wisdom of deploying Indonesian or Egyptian troops to combat Muslim extremists who are likely to have their own sympathizers within those forces—and who are certain to work hard to infiltrate them and gain new adherents.

As America's friends and allies are disarming, our adversaries are arming. The Chinese have been engaged in a significant expansion of their military power for years. The Russians have also declared their intention to upgrade their nuclear arsenal, if nothing else— and they have the oil resources to expand their conventional forces as well if they chose. The Iranians have been expanding the reach and power of their terrorist portfolio in the Middle East. Even Venezuela has embarked on an expansion of its armed forces, although it is unlikely to be a meaningful threat outside its immediate environs, whatever the course of that program, any time soon. These trend lines are most worrisome. America's current and potential foes are expanding their capabilities across the spectrum from nuclear war to terrorism even as our allies continue to reduce their military expenditures. The United States has demonstrated an astounding inflexibility in its own response to changing threats and international conditions. And the U.S. military posture now contains completely unacceptable levels of risk.

The Nature of the War on Terror

The entire discussion about the future of the American military is further confused by the complexity of the idea of a "global war on terror"(GWOT) and how to fight it. President Bush coined the term almost immediately after the 9/11 attacks, and has prosecuted this war aggressively with both conventional campaigns (in Iraq and Afghanistan) and unconventional measures (including Special Forces but also financial, informational, and diplomatic efforts). Right now, the GWOT is very close to being a full-spectrum conflict—only nuclear weapons remain unused, although their deterrence value remains high with regard to Iran. Indeed, the term "Long War" better captures the nature of this conflict.

Many have argued that this approach to the struggle against terrorism is mistaken. The United States, they maintain, should abandon its high-end conventional operations and focus exclusively on low-intensity efforts, relying militarily on Special Forces and making greater use of diplomacy and other nonmilitary means to gain local allies. Opponents of the current strategy argue that the use of conventional forces is counterproductive, that it makes more enemies than it kills, and that it should be abjured.[15] They sometimes claim that American conventional forces should be reduced both to eliminate the temptation to prosecute the wrong kinds of wars and also to make resources available to increase Special Forces, foreign aid, and other unconventional means of waging an unusual kind of war.

It has been the purpose of this and the preceding chapter to show that the United States has more enemies than just al Qaeda—to say nothing of threats, challenges, and requirements. The U.S. armed forces cannot be optimized simply to fight the Long War, however important that struggle is, without regard for other, more traditional and conventional threats and dangers. Even if the right force mix for this war excluded high-end conventional forces, that mix will assuredly not suffice against China, is unlikely to succeed against Iran, and does not provide adequate safeguards for collapse scenarios in Pakistan, North Korea, or Saudi Arabia.

But in fact there is little evidence from the last few decades to suggest that low-intensity, unconventional, or over-the-horizon approaches are likely to succeed against al Qaeda and similar movements. The United States did not intervene in Afghanistan or Somalia in the 1990s to prevent the triumph of groups linked to al Qaeda, and those groups (the Taliban and the United Islamic Courts) ultimately took power. Their seizures of power did not come in response to ham-handed American occupation. In the case of Somalia, they came despite a concerted effort by Joint Task Force–Horn of Africa to use Special Forces and other forms of soft power to prevent their triumph while maintaining a very low profile and presence. Both the Islamic Courts and the Taliban were removed from power, on the other hand, by conventional attacks— indigenous Afghan forces supported by U.S. airpower in the one case, and a simple invasion by the Ethiopian Army in the other.

One might argue that these events were proof of another principle— that it is better to use local or indigenous forces than American troops—but that conclusion is also problematic. The exclusive reliance on Afghan forces with varying agendas and no U.S. control on the ground compromised subsequent American efforts to help the fledgling Afghan government gain control over the historically wayward provinces—U.S. money had helped strengthen the very warlords that Hamid Karzai then had to face down. In the case of Somalia, the Ethiopian Army is clearly incapable of maintaining its occupation, even were that desirable, and there does not appear to be any other force on the horizon ready to take its place. Success against the Islamic Courts may prove illusory because of the weakness of the regional proxy that helped pursue American interests in the region.

Pakistan is the glaring and outstanding example of the failure of this theory of how to fight the Long War. U.S. forces have studiously avoided large-scale incursions into the tribal areas of Waziristan and Baluchistan that now harbor al Qaeda bases and leaders, and American leaders have worked instead to use diplomatic and other soft power levers to convince their Pakistani allies to rein in the terrorists in their midst. These efforts have had little effect. The Pakistani military has proved unwilling or unable to take on al Qaeda; both al Qaeda and a resurgent Taliban (probably supported by elements of the Pakistani government) are growing in their ability to undermine the Karzai government in Kabul; and fears of destabilizing Pakistan's weak secular government seem to have paralyzed Washington. It is clearly inappropriate to conclude from this example simply that the United States should invade Pakistan. But it is very hard to look at any of these examples and conclude that soft power and Special Forces–style raids or support to indigenous governments is likely to be a successful way to prosecute the Long War.

And the cases of Afghanistan and Iraq offer far more complicated lessons about the best ways to fight al Qaeda than common wisdom in Washington often holds. Before November 2001, al Qaeda had large-scale permanent bases and training camps in Afghanistan from which it was able to plan and launch global terrorist operations. After not merely the fall of Kabul, but the insertion of an American division into the country, there were no large permanent al Qaeda bases in Afghanistan, although the situation is deteriorating again following a premature effort to reduce the U.S. presence and transition responsibility for the counterinsurgency effort to NATO.

Al Qaeda did not have large permanent bases in Iraq before the U.S. invasion, but it established them rapidly in the Sunni Arab areas after the fall of Baghdad with the support of the Sunni Arab community. U.S. strategy in Iraq that focused on keeping our footprint small and working through the nascent Iraqi Security Forces rather than undertaking our own clearing operations allowed al Qaeda in Iraq (AQI) to grow and metastasize—to gain additional and more secure bases not only in Anbar, but around Baghdad and into Babil, Ninewah, Diyala, and Salah-ad-Din provinces. A shift in the American approach combined with al Qaeda errors led to a sea change in that situation in 2007. U.S. forces began in mid-2006 unilaterally to clear Ramadi and other AQI bases in Anbar, and the Sunni Arab population began to turn against al Qaeda not only in Anbar, but throughout Iraq. At the moment of this writing,

U.S. and Iraqi forces are engaged in major operations against al Qaeda bases throughout Iraq, and have been receiving help ranging from tips to troops from local Sunni Arab communities that until recently had supported, or at least tolerated, the terrorists.[16]

This process is far from complete, of course, and it is not necessarily replicable everywhere al Qaeda has bases. It does seriously call into question, however, the shibboleth that the use of American conventional forces to clear terrorists out of bases is inherently and hopelessly counterproductive. Looking at the events of the past year in Anbar, and of the past seven years in Afghanistan, Pakistan, and elsewhere, it is not at all clear that low-profile, over-the-horizon operations are the preferred means of fighting al Qaeda. There is certainly not enough evidence of that thesis to justify reorienting the entire American military on that single approach to this struggle. On the contrary, recent experience suggests, again, the need for a full-spectrum force able to wage every level of war from the highest to the lowest, and to do so discretely and intelligently within a single theater and across theaters. There is clearly no one-size-fits-all approach to the war on terror, any more than there is to any large and complex global struggle. The U.S. military will certainly need all the tools available to defeat the challenge of Islamic terrorism.

The Restoration of Military Capability

The United States must undertake an urgent program to restore its overall military capability to a level at which it can face visible and probable threats on its own with equanimity in the near and middle term. Because of the heavy burden that current and likely future threats place on the ground forces, this program must build America's ability to fight high-end conventional, mid-range counterinsurgency, and low-end counterterrorism threats simultaneously without placing intolerable stress on the military. America should undertake such a program at once, and push it as rapidly as possible.

The Iraq War provides a convenient cloak for rebuilding the ground forces. Since the United States is not openly fighting or preparing to fight Iran, and since the strains the Iraq War place on the army and marines are so well known and documented, the expansion of those forces now would seem natural (indeed, the failure to expand them to date has seemed highly unnatural). If we wait until conflict with Iran appears imminent before beginning to build our military capabilities, then the simple act of military expansion might accelerate a growing crisis. Conversely, the fear of accelerating a crisis may well play a powerful role in delaying or stopping the necessary expansion of our ground force capabilities. Escalation of the conflict with Iran is not desirable if it can be avoided without compromising vital U.S. interests. But it is essential for the United States to have the capability to prevail in such an escalation if it becomes necessary. The best time to get ready for this or any other dangerous scenario is before the crisis is upon us.

3

Case Studies: New Battlefields

Since the end of the Cold War—and the disintegration of the Red Army that so drove the development of U.S. land forces and the U.S. Army in particular—the number and variety of military operations on land has mushroomed. Just six weeks after the fall of the Berlin Wall, President George H.W. Bush gave the green light to Operation Just Cause, the lightning strike that toppled Panamanian strongman Manuel Noriega. In retrospect, this miniwar revealed many of the elements that have characterized subsequent conflicts. The conventional combat was the sort of large-scale *coup de main* that became the ideal of the defense transformation movement: airborne and special operations forces were deployed rapidly from their bases in the United States; twenty-seven separate targets were "taken down," including the main headquarters of the Panama Defense Force, in eight hours; Noriega himself escaped but was eventually taken. Less noticed was the fact that half the initial force was already stationed in Panama; much of the fighting took place within sight of the headquarters of U.S. Southern Command. Almost entirely overlooked was the postcombat chaos. Panamanians took to the streets not to defend the regime or to protest a *Yanqui* intervention, but to loot their own shops; overcoming the initial resistance of their commanders and political leaders, American troops were soon devoted to police functions lest the victory won so quickly on the streets of Panama City be lost in the press coverage.[1]

The 1990s proved to be a violent decade, despite the general great-power peace. As then director of Central Intelligence James Wolsey summed it up: the United States and its allies had, in the Cold War, slain the Soviet dragon only to find the "jungle filled with a bewildering variety of poisonous snakes."[2] Freed from the constraints of the bipolar, superpower standoff, petty tyrants aspiring to smaller dreams of hegemony sought to find the seams in Bush's emerging "New World Order." First out of the gate, fittingly, was Saddam Hussein; the 1990 invasion of Kuwait was a continuation of the impulse behind the long, bloody Iran-Iraq war.[3] The American-led war to drive Saddam back to Iraq was decided ultimately by the immense power of the Desert Storm land force, comprising five hundred thousand soldiers and marines employing more than two thousand M1 tanks and other armored combat vehicles. The conflict revolved around a number of short, sharp—and profoundly one-sided—tank-on-tank engagements reminiscent of Rommel in North Africa, Patton in France, or the crushing Soviet tank armies of the eastern front in World War II. But again, if the battles were decisive, the war was not: Saddam's Iraq was a constant menace that demanded daily American military attention and the continuous rotation of land forces to the region.

Nor did Saddam's defeat seem to have much of a deterrent effect elsewhere. When the humanitarian operation begun by Bush in Somalia became an open struggle for power in Mogadishu under his successor Bill Clinton, what was a temporary deployment of marines became an extended deployment of a reinforced division of American soldiers, a large contingent of special operations forces, and a rag-tag international coalition.[4] Deepening involvement in Somali politics led to a deepened military commitment, but when the infamous "Black Hawk Down" events of October 1993 proved a tremendous domestic political embarrassment—and led to the resignation of Defense Secretary Les Aspin—Americans seemed to lose their appetite for land wars in faraway nations with fractious peoples. Rwandans and Bosnians suffered in part because the United States had become gun-shy; any impulses toward intervention felt in the Clinton administration were strongly checked by the new Republican opposition majority in Congress. But eventually the embarrassments of Slobodan Milosevic's ethnic cleansing campaigns in the Balkans became greater than the potential embarrassments of defeat. The U.S. First Armored Division led the NATO

Stabilization Force in Bosnia in 1996, and in 1999 America and its European allies expanded their presence in the former Yugoslavia by occupying Kosovo. These were not high-intensity combat operations—although the unquestioned dominance of U.S. land forces was a key to keeping an otherwise uneasy peace—but reintroduced "nation-building" to the lexicon of the American military.[5]

Since the attacks of September 11, 2001, the intensity, diversity, and pace of U.S. land force operations have increased exponentially. Indeed, it seems that the blur of battle has become so great that we have yet to begin to analyze what, precisely, has been our recent experience. And what follows, in the brief discussions of five illustrative post-9/11 ground campaigns, can only suggest the outlines of this experience; if we are to understand the overall, institutional requirements for American land power, a more comprehensive set of studies is needed. Nonetheless, what emerges from these quick case studies is a sense of the qualities of modern combat across a wide spectrum of operations. The United States, it is clear, must have land forces capable of fighting not only at the times, places, and manners of our choosing, but in response to durable, adaptable, and lethal enemies.

The Invasion of Iraq: Speed Kills

The difficulties of the counterinsurgency campaign in Iraq have all but obscured the remarkable success of the initial invasion. In less than three weeks, from March 20 to April 9, 2003, U.S. and coalition ground forces entered Iraq, fought their way into the heart of the country, captured Baghdad, and removed Saddam Hussein's regime from power. The operation underscored the ability of the United States to conduct, along with the British Army, combined-arms, air-land warfare at an unprecedented level of professionalism. Despite self-imposed planning restrictions and a mismanaged deployment that rendered the largest and most modern U.S. formation unavailable at the war's outset, the superior quality of the invasion force more than sufficed to topple Saddam Hussein's regime at a minimal cost in blood and treasure.[6] The confidence of superiority played a large role in allowing commanders to think boldly, deploy rapidly, and, in fluid situations, seize fleeting initiatives or respond to unanticipated developments.

Prewar planning was based upon a complex maneuver resulting in a converging attack on Baghdad. The coalition forces were to advance on the Iraqi capital mainly from the south, along three axes. The main thrust was to be delivered on the left, by a U.S. armored division advancing deep into Iraq on the west bank of the Euphrates River before crossing the river and advancing on the city itself through the Karbala Gap. The First Marine Expeditionary Force was to provide a supporting punch on the right, advancing from Nasiriyah on the Euphrates to Kut on the Tigris, and then from Kut to Baghdad along the east bank of the Tigris River. A third, smaller force would launch a holding attack in the center, advancing between the two great rivers, with the purpose of tying down Iraqi forces. To top it all off, airborne forces would land to the north of the Iraqi capital, cordoning off the city and acting as the anvil to the southern columns' hammer. In addition to the forces advancing from the south, a fourth thrust was to converge on Baghdad from the north, passing through, in turn, Turkey, the autonomous Kurdish region of Iraq, and the Sunni heartland north of Baghdad. In sum, the plan had a lot of moving pieces, relied upon the synchronization of many maneuvers, and assumed tactical supremacy at the point of contact with the enemy. A further notable difference from the 1991 Desert Storm land campaign was the lack of a long preparatory air campaign.

The actual invasion differed significantly from the prewar plan, but in ways that accepted greater risk. To begin with, the thrust from the north never happened. Turkey refused to allow coalition forces to use its territory, precluding the planned northern thrust. Embarked on ships off the Turkish coast at the start of the war, the U.S. Fourth Infantry Division (the U.S. Army's most heavily armed and modernized division) was forced to sail to the Persian Gulf. Missing the first two weeks of the war, it acted as a follow-on force to the forces which invaded Iraq from the south at the beginning of the war. Furthermore, political considerations prevented the coalition from massing all the desired forces (such as the army's logistics and bridging units, vested in the reserves and National Guard) in Kuwait prior to the commencement of the invasion. In Secretary Rumsfeld's judgment, activating the reserves and National Guard and massing the full-sized invasion force in the region while negotiations at the UN were still underway would have made the United States' efforts at negotiating look disingenuous.[7] Once negotiations had failed, waiting to commence operations while the invasion force massed would have been politically undesirable as well. The decision was made to send in "force packages" as they became available—Pentagonese for attacking in a piecemeal fashion rather than massing as much force as possible. As a consequence of this decision, the planned holding attack between the Euphrates and Tigris rivers was abandoned for want of troops.

Despite these deviations from prewar planning, the invasion was overwhelmingly successful from the beginning. At the start of the campaign the Third Infantry Division quickly captured Nasiriyah, a vital crossing point on the Euphrates River, which helped speed up the marines' advance in the east considerably. Meanwhile, the marines quickly achieved their initial objective of seizing the Az Zubayr oil pumping station intact. East of the marines, the British First Armored Division seized the al Faw peninsula and cordoned off Basra, Iraq's second-largest city.

The second phase of the invasion was highly successful as well. West of the Euphrates, the Third Infantry Division quickly advanced from Nasiriyah to the western side of the Karbala Gap, handily pushing aside such feeble conventional resistance as it encountered. At the same time, the pace of the advance taxed logistics units and rendered them vulnerable; the line of communication and supply back to Kuwait was long and thin. While the Third Infantry Division quickly reached Karbala, its rear areas came under attack from the Fedayeen Saddam, fanatical Baathist irregulars. As the mechanized forces lacked sufficient forces to both secure their lines of communication and punch through the Karbala Gap to Baghdad, the reserve of airborne infantry, including the highly mobile 101st Airborne that had been slated to leap ahead to cut off Baghdad, was committed during the second week in order to secure the tenuous thread of logistics. Meanwhile the marines made excellent progress driving towards their intermediate objectives of Diwaniyah and Kut. Also, as a makeshift substitute for the Fourth Infantry Division's thwarted northern thrust through Turkey, a small force consisting of Special Forces, paratroopers from the 173rd Airborne Brigade, and a small armored unit was inserted by air into northern Iraq to open up a second front, acting in concert with local Kurdish *peshmerga* militia. These forces operated with considerable success in northern Iraq, but lacked the necessary weight to drive south into the heart of the country. Despite the unpredicted ferocity of the irregular attacks, the invasion force reached a position to begin the final drive toward Baghdad with incredible speed and without encountering any large-scale resistance.

The final thrust was delayed by a heavy sandstorm which began on March 25 and lasted until March 29. During this pause, however, the coalition continued to bombard the Iraqi Army from the air and work on securing its lines of communication; perhaps most importantly, the U.S. ground forces set themselves to deliver the blow to Baghdad. When the weather cleared, the Third Infantry Division sprinted through the Karbala Gap—the constricted western approach to Baghdad and the place where U.S. planners had long feared Saddam had his best opportunity to employ chemical weapons. Here, too, was the most likely position for the Iraqi Republican Guard to mass in defense, but the air attacks, unconstrained by the weather, had crippled the Iraqi efforts, and the U.S. Army forces swept through relatively light and poorly organized resistance. The marines were equally successful in the east, breaking through a Republican Guard defensive line at Qalat Sikar. By April 3, the stage was set for the battle of Baghdad.

This final phase of the campaign—the attack into the city itself—had been the fight most feared by U.S. commanders. As pundits never tired of pointing out, urban combat can be lethal and chaotic; the ghosts of Stalingrad were repeatedly summoned. Yet they never came. On April 3, the Third Infantry Division began its assault on Saddam International Airport, securing a foothold that day and the entire airport the next; the marines pressed the capital from the south. On April 5, the Third Infantry Division launched the first of what it called "thunder runs," a raid by a force of thirty M1 tanks and fourteen Bradley fighting vehicles through the streets of Baghdad.[8] The raid aimed to demonstrate the impotence of Saddam's regime by driving right past Iraqi ministries and other symbols of the regime's authority. Supported by air force A-10s and army attack helicopters, the first thunder run was extremely successful. The column easily completed its circuit, leaving mounds of dead Iraqi defenders in its wake and losing only a single tank in the process.

Buoyed by the success of the operation, the Third Infantry Division launched a second, larger thunder run on April 7. This column seized Saddam's presidential palace and held it. Also on April 7, the marines attacked the southeast quadrant of the city. These operations spelled the end for the Saddam regime; Iraqi forces organized a desperate counterattack that night, but when it failed, it precipitated a general collapse of the Iraqi army. By April 9, organized resistance in the Iraqi capital had ceased.

The conventional mission was indeed accomplished. Even allowing for the many inherent weaknesses of the Iraqi Army, the pounding it had received in Desert Storm, and the pressures maintained in the subsequent years, the invasion campaign achieved an extraordinary and rapid victory with an absolute minimum of forces. It is this capacity for combined-arms, air-land, joint, and coalition warfare that distinguishes U.S. land forces; this is what they can do that no other forces can.

Among the many factors that create these overwhelming tactical and operational advantages, two stand out in analyzing the Iraq invasion. First, the campaign was built upon the decisive advantages enjoyed in direct-fire engagements that come from the quality of U.S. land combat vehicles, most notably the Bradley infantry fighting vehicle and the M1 Abrams series tanks. The second is the unprecedented ability of U.S. forces to integrate land power and airpower, providing American land maneuver formations an unprecedented level of fire support; the flexibility, responsiveness, and precision of U.S. air forces make them almost an extension of the direct-fire engagement.

Heavy armored vehicles, and especially the main battle tank, often are dismissed as Cold War relics. But modern land combat systems are immensely more sophisticated than their predecessors; they retain the traditional qualities of firepower and mobile protection, now magnified by the thermal sights and on-board computers that make an M1 not only a lethal platform in itself but an important node in a networked force. In addition to handily defeating Iraqi armored units in open country, as they had done in 1991, U.S. heavy forces—again, the M1 series tanks stand out—proved their worth in an unexpected setting: the fight for Baghdad. The two successful "thunder runs" were highly unorthodox, as historically tanks have been highly vulnerable to enemy infantry and anti-tank guns when fighting in an urban environment. And though these attacks, initially by small U.S. units, met fanatical, even suicidal resistance—nearly every tank was hit by enemy fire, many by multiple rocket-propelled grenades—they lost only a single vehicle. The manifest inability of the Iraqis to destroy American armor, even in an urban environment, perhaps the most favorable circumstance in which to do so, effectively ended the conventional phase of the Second Gulf War by demonstrating the futility of organized resistance. It also spared the American infantry and marines the difficult and inevitably bloody task of rooting Iraqi soldiers out of every building and sewer line in the teeming Iraqi capital. The after-action report of the Third Infantry Division recognizes the importance of the M1 Abrams' survivability: "This war was won in large measure because the enemy could not achieve effects against our armored fighting vehicles."[9]

The second significant factor in the success of the invasion was the quantity and quality of airpower, both in direct air support and in air interdiction. While the "shock and awe" bombing campaign against Baghdad at the beginning of the war was the best publicized employment of airpower, the most significant military contribution by the coalition's aerial assets was its excellent support of the invading ground forces. Broadly speaking, the coalition's air assets supported the ground forces in three critical ways: softening up Iraqi units before they were engaged by coalition ground forces, interdicting Iraqi troop movements, and providing fire support at the point of battle.

The coalition's aerial assets did an excellent job softening up the Iraqi units in the path of the coalition's spearheads, especially the Republican Guard armored units. The greatest demonstration of aerial preparatory bombardment was the battle for the Karbala Gap. As discussed above, the Karbala Gap, a three-mile–wide choke point, was the natural place for the Iraqis to make a stand to check the coalition's drive on Baghdad. While the Third Infantry paused outside the gap, waiting for a sandstorm to clear, the coalition's aerial assets continued pounding the Republican Guard units that might have moved into the area. The intensity of the bombardment was such that the Iraqis were unable to mass their armored forces in or near the gap to resist the inevitable coalition attack. When the sandstorm cleared and the ground forces attacked, they did not meet any organized resistance until they were well past the gap. The coalition's aerial bombardment had denied the Republican Guard the opportunity to fight at the best location on the Third Infantry Division's route to Baghdad.

The coalition's air forces also did an excellent job of providing fire support once the ground forces had met the Iraqis in battle. The importance of the coalition's close air support was demonstrated in the battle for Kifl, a village and crossing point on the Euphrates River. The bridge was seized by an infantry company of the 101st Airborne, which established itself on the east bank of the river. This advance element was exposed when a sandstorm prevented the division from immediately reinforcing the lone company with additional ground forces. The division's attack helicopters protected the stranded and outnumbered company until the weather allowed the division to reinforce them.

In sum, the invasion of Iraq is a reminder of the requirements for victory in large-scale conventional campaigns, and of the utility of U.S. supremacy in this kind of land combat. The ability of the United States to topple threatening regimes remains a powerful force in international politics. However we find ourselves chastened by the challenges of irregular warfare in Iraq and Afghanistan, the invasion of Iraq provides a standard for measuring the conventional capabilities—to maneuver large forces rapidly over large distances, to close with the enemy and dominate the direct-fire engagement, to coordinate airpower and land power—that remain enduring prerequisites for American land forces.

Tal Afar: Conventional Forces in Irregular War

The first phase of the American occupation in Iraq, commencing shortly after the fall of Baghdad in early April 2003 and continuing through most of 2005, proved to be disastrously chaotic for the coalition. Personal relations between L. Paul Bremer, the U.S. ambassador and administrator of the Coalition Provisional Authority, and his military counterpart, Lt. Gen. Ricardo Sanchez, were terrible, and, despite the increasing signs of counterinsurgency, there was too little strategic coordination as well.

The second phase featured improved civil-military cooperation and policy coordination between Ambassador Zalmay Khalilzhad and Gen. George Casey. But their "train and transition" approach, based on faulty assumptions of Iraqi capability, began to collapse when, in early 2006, al Qaeda in Iraq finally succeeded in provoking the large-scale sectarian conflict that long had been its stated aim.

The third phase—the combination of classic counterinsurgency "protect the population" tactics and increased American presence popularly known as "the surge"—seeks to remedy the failures of the plans that went before. But ironically, the model for the surge was a product of the first, most chaotic period. It was the remarkable success of the Third Armored Cavalry Regiment (ACR) in the town of Tal Afar in far northwest Iraq that provided the blueprint for the later Baghdad Security Plan.

Indeed, the extended battle of Tal Afar underscored the ability— also the necessity—of conventional units to achieve significant and lasting successes in irregular warfare. When Colonel H. R

McMaster's Third ACR, numbering five thousand troops, arrived at Tal Afar in the late spring of 2005, the city of two hundred thousand inhabitants in the northwest Ninewa province was thought to be "the next Fallujah"—a reference to the bloody fighting by the marines the previous year in that city of Anbar province. In Tal Afar, Iraqi authorities had lost control, and al Qaeda insurgents or similar *takfiri* Sunni extremists were terrorizing the city's population. It was also an important initial way station for terrorists and suicide bombers infiltrating from Syria.

These challenges had thus far posed problems that conventional coalition forces had been unable to solve. In fact, in many cases, American tactics were making things worse. The basic approach to achieving stability at that point was to find and kill the insurgents or, in an extreme emergency like Fallujah, to conduct large-scale cordon-and-clear operations. But McMaster broke with this convention, and was the first to prioritize protecting the population over simply killing the enemy's forces. This transformation was all the more remarkable for taking place in a cavalry regiment, a unit designed first and foremost to maximize mobility and firepower, not to hold ground in a sustained way.

Back at Fort Carson, Colorado, members of the regiment had undergone a rapid introduction to the Arabic language and Iraqi culture to orient themselves to their challenges on the battlefront. Having prepared for many scenarios where cultural awareness would be useful, McMaster and his deputy, Lt. Col. Chris Hickey, knew how and where to begin their work in Tal Afar. Coalition troops first cleared neighborhoods of insurgents, engaging in frequent combat. However, instead of disengaging after killing the enemy, troops stayed to "hold" the neighborhood, emphasizing tactics to protect the population.

A small unit in a large town, the regiment understood the necessity of establishing relations with the Iraqis; the residents of Tal Afar had to embrace the American presence and view U.S. soldiers not as occupiers but as their defenders. This is where the special training repaid the regiment's efforts: soldiers learned and participated in local customs, and earned the trust of those whom they were protecting, usually Shia under attack from fanatical Sunnis. By the end of the operation, reported George Packer in *The New Yorker*, the residents knew the American soldiers by name; some soldiers even brought gifts for the Iraqi children when they were leaving.[10]

Besides the action of individual platoons in clearing and holding, the regiment's senior officers were also involved in establishing ties and gauging the situation better. McMaster, with the help of an attachment of three thousand well-trained Iraqi soldiers, was able to establish outposts outside the perimeter of Tal Afar and construct a reliable intelligence network. Again, this was a profound departure from previous and much subsequent U.S. practice; the Third ACR treated Iraqi soldiers as allies, not auxiliaries. The regiment's deputy, Lt. Col. Hickey, spent a large amount of his time deliberating with the Turkmen, Shia, and Sunni tribes in order to arrive at some understanding. He was also instrumental in reorganizing the Tal Afar police force. These efforts began during the initial clearing and holding phases of the operation, and contributed in the final and most difficult phase, the "building" of long-term security and social, political, and economic reconstruction. As part of the concerted "building" effort, coalition troops helped install sewage and water facilities and rebuilt shops destroyed by weeks of firefights; rationed food was distributed by American and Iraqi troops. By September 2005, the state of affairs in Tal Afar had considerably improved.[11]

The decision to hold on to the cleared neighborhoods of Tal Afar also created improved conditions for follow-on clearing and combat operations. In particular, intelligence networks developed by senior officers were aided by local intelligence from Iraqis who had learned to trust the Americans: once American soldiers had proven to Iraqis that their primary responsibility was protecting them, not killing them, the two sides could establish a rapport, and the Iraqis began slowly to share information about the whereabouts of insurgents and the locations of roadside bombs. Sunni and Shia leaders, under supervision of Lt. Col. Hickey, collaborated to bring an end to the anarchy that had gripped Tal Afar. In October 2005, Secretary Rice announced the army's new strategy of "clear, hold and build," pointing to Tal Afar as a success story.[12] In March of the following year, President George W. Bush cited Tal Afar as an indicator of possible success in Iraq.[13]

Three factors allowed the coalition under McMaster to secure Tal Afar. The first relates to the size of the force used. A significant reason for the coalition's failure to maintain Tal Afar after the 2003 invasion was its meager troop presence: CENTCOM left only five-hundred-odd American troops in the city. These Americans, along with the Iraqis (who were, in all likelihood, not yet well trained), were clearly overwhelmed by the Sunni insurgents, who, in turn, were probably aided by most of the 150,000 Sunnis in the city.

Col. McMaster's operation to clear, hold, and build Tal Afar, on the other hand, was the largest offensive in Iraq since the initial invasion in March 2003. Still, it was an operation that both confounded and confirmed aspects of conventional wisdom about the ratio of troops to population in counterinsurgency operations. McMaster used about five thousand U.S. troops to secure a population of roughly two hundred thousand persons. While there were about three thousand Iraqi soldiers, their relatively inferior training made them less capable of taking the lead in counterinsurgency operations. They served superbly in their function of providing a cultural interface and in the hold-and-build operations, but were less involved in the clearing operation. Thus, in practice, there was one U.S. soldier for every forty inhabitants of the city—enough to provide security but few enough to avoid appearing as invaders.[14] It is important to note that the residents of Tal Afar did not complain about the Americans acting as colonizers; Iraqis in Tal Afar respected not just the Americans' aims, but also their endurance in the face of difficult odds.

The second important factor that proved vital in securing Tal Afar was the Third ACR's relationship with Iraqi troops, facilitated by its own level of training in the cultural sphere. Without the necessary cultural interface, McMaster would have been able to accomplish nothing. The Iraqis' role in the reconstruction efforts is probably larger than that of their American counterparts, considering that they possessed greater local knowledge than the American troops. The cavalry regiment's ability to forge this relationship—at a time when the overall program of training Iraqi forces was being reorganized—makes clear that it is possible and indeed necessary to employ conventional units as well as Special Forces in building partnership capacity. The preparations at Fort Carson did not make the soldiers of the Third ACR into highly qualified Arab linguists or scholars, but they did provide an essential awareness of the local culture and language. These soldiers were members of a highly conventional unit, but they were well prepared by their leaders for a mission that demanded something more than fire and maneuver. The lessons of Tal Afar have now been codified in the new joint army and marine counterinsurgency doctrine. As Gen. David Petraeus explains:

> Cultural awareness is a force multiplier. It reflects our recognition that knowledge of the cultural "terrain" can be as important as, and sometimes even more important than, knowledge of the geographic terrain. This observation acknowledges that the people are, in many respects, the decisive terrain, and that we must study that terrain in the same way that we have always studied the geographic terrain . . . Understanding of such cultural aspects is essential if one is to help the people build stable political, social, and economic institutions.[15]

The third factor that led to success at Tal Afar was the greater role of leadership, specifically, the ability of leaders to innovate in gauging and solving problems, and to risk deviating from the conventional overarching strategy. While it is true that McMaster and his lieutenants introduced a panoply of clever tactics, they did more: in effect, they reoriented American military strategy, at least in Tal Afar. The prevalent approach to the growing security problems in Iraq in 2004 and 2005 did not acknowledge the conflict as an insurgency. Moreover, it was the belief of Gen. John Abizaid (under whom McMaster had previously worked) that Americans inevitably provoked deep xenophobia among Iraqis and that U.S. units should have a small "footprint" and avoid a strongly interventionist approach. By ignoring these two premises, McMaster was able to secure Tal Afar. The resulting contrast between what the regiment was able to accomplish in Tal Afar, and the transitory effect of the marines' hard-won victory in Fallujah, is striking; in Iraq, it is clear, the "hold" phase is often the decisive form of combat, setting and maintaining the conditions for the "build" phase.

While many of the lessons of Tal Afar might seem to be operational, there are institutional implications as well. The need for specialized training prior to deployment and for sufficient force ratios, for example, is directly correlated to the overall size of U.S. land forces. Proper training means time, people, and money. But an equally important, if somewhat less obvious, lesson is the ability of properly trained, led, and resourced conventional units to succeed in irregular warfare missions. Both the army and Marine Corps face a wide variety of missions, and the temptation to create special-purpose units, not only traditional special operations forces but less traditional formations such as "peacekeeping brigades," is strong, amounting in some quarters to a fad or a fetish. And of course some missions are beyond the ability of conventional units. But the success of the Third Armored Cavalry Regiment demonstrates the flexibility of "general purpose forces" in unexpected environments; a variety of timeless characteristics, such as superior leadership, tailored and mission-relevant training, unit cohesion, and the exercise of tactical initiative, were the underlying factors that led to the successes in Tal Afar.

Tal Afar raises crucial questions as the land forces of the United States expand and change to meet the challenges of the Long War. Of course the army and Marine Corps must continue to produce tactical competence, units that have mastered the elusive arts of fire and maneuver in combat. But what next? What other skills must leaders possess? Col. McMaster is not an expert in Middle Eastern history or languages. He is, however, a distinguished historian and student of war. He knew what skills his regiment would need to succeed in a counterinsurgency in Tal Afar; he knew how to mold his unit for its mission; he understood, even before his superiors did, the political and strategic framework in which he would be operating. What seems clear in retrospect is that McMaster's serious education and training in the higher arts of war served him—and his regiment—more than a similar effort studying Arabic.

Israel in Lebanon: Serial Surprise

The last generation of conflicts in the Greater Middle East has involved "serial" wars; that is, sustained campaigning may be infrequent and episodic rather than constant, and the underlying political conditions change only very slowly. The U.S. involvement with Iraq certainly fits such a pattern; dealing with Saddam Hussein was a drama of many acts, and it is hard to see the current effort at reconstruction reaching a conclusion—positive or negative—in the near future. It is not unreasonable to view the American experience in Afghanistan similarly, from the efforts of the 1980s to support the anti-Soviet *mujahideen* through the Taliban years to today. But the condition of serial war is, for Israel, a strategic fact of life on multiple fronts.

Thus it is best to consider Israel's 2006 Lebanon War as a continuation—after a six-year interlude—of a much longer conflict that began in the 1980s. In 1982, Israel, seeking to prevent attacks by Palestinian militants operating out of Lebanese territory, invaded Lebanon. The terrorist organization Hizbollah was founded in response to resist the Israeli occupation.[16] In the 1980s and early 1990s it conducted a largely unsophisticated campaign, relying mainly on suicide bombers and other rudimentary terrorist methods. In the mid-to-late 1990s, its organization improved, enabling it to fight a guerrilla war against Israel in southern Lebanon (as opposed to carrying out isolated acts of terrorism).[17] In 2000, Israel withdrew from southern Lebanon, seeking to avoid the cost in lives and treasure of a seemingly endless guerrilla war on its northern front while it was already fighting a similar conflict with the Palestinians—and worrying about Iran's growing nuclear capabilities. Hizbollah, needing to prolong the conflict for its own strategic reasons and to validate its role in Lebanon, cited Israel's continued occupation of the Shebaa Farms as its *casus belli*,[18] and continued harassing Israeli troops on the border while fortifying southern Lebanon in anticipation of another Israeli invasion.[19] While Israel had to turn its attention in other directions, Hizbollah looked forward to the next installment in the serial war.

During the six years which followed Israel's withdrawal from Lebanon, Hizbollah's military capabilities increased considerably, continuing the trend which began in the 1990s. Its organization tightened, and it received advanced weapons (such as modern anti-tank missiles and electronic warfare technology), as well as training in their use, from Iran and Syria. The relative quiet (save for some desultory skirmishing on the border) gave Hizbollah the breathing space to absorb the new technology and tactics. Additionally, it acquired a large arsenal of Russian, Syrian, and Iranian surface-to-surface missiles, giving it the ability to attack population centers in northern Israel. Just as important was Hizbollah's increasingly strong political grip on southern Lebanon, which legitimized its military presence and permitted the construction of many defensive strongpoints that were well camouflaged and integrated with civilian structures. All in all, Hizbollah was a much deadlier and more sophisticated force in 2006 than it had been when the Israeli Defense Forces (IDF) left southern Lebanon in 2000.

In July 2006 Hizbollah raised the stakes by launching a carefully planned raid into northern Israel, killing eight Israeli soldiers and capturing two. While Hizbollah probably did not yet intend to provoke a broader war—it intended to swap the captured Israeli soldiers for Arab prisoners held by Israel—the preparations of the preceding years had essentially increased Hizbollah's readiness and its capacity to withstand a strategic surprise.[20] Israel refused to return the Arab prisoners and began a military campaign with the declared objectives of disarming Hizbollah—especially its rocket arsenal—and securing the return of its two soldiers. Hizbollah's objectives on the other hand were far more limited: its leader Nasrallah declared that victory would be achieved if Hizbollah survived the Israeli onslaught.[21]

Israel pursued these rather ambitious strategic goals with decidedly limited means: its initial military strategy might have been

lifted from Donald Rumsfeld's vision of transformational strike warfare, including a massive bombing campaign against Hizbollah's
military assets, leadership, and the infrastructure of Lebanon at
large, accompanied by limited numbers of special operations forces
on the ground to assist with target acquisition. Hizbollah was neither shocked nor awed and responded by augmenting its barrage of
surface-to-surface rockets on northern Israel. Israel's air campaign
was supposed to disarm Hizbollah both directly, by destroying its
military assets, and indirectly, by turning the rest of the population
against Hizbollah, making its position untenable. It was successful
in neither regard. While the Israeli Air Force was able to destroy
most of Hizbollah's long-range missiles at the beginning of the conflict and eventually figured out how to detect and destroy its
medium-range missiles, it was unable to destroy its arsenal of short-
range missiles, which constituted the vast majority of Hizbollah's
capacity. The desired indirect effect of weakening popular support
for Hizbollah failed miserably at both the local and the international
level.[22] As Israeli bombs rained down on Lebanon's infrastructure,
the Lebanese took a more immediate view and coalesced against
Israel rather than blaming Hizbollah for provoking the Israeli
onslaught. Even Hizbollah's most bitter and longstanding political
opponents took its side in the crisis—some even taking up arms
against Israel in southern Lebanon. At the international level, the
bombing campaign, perceived around the world as indiscriminate,
ineffective, and disproportionate, turned world opinion almost
unanimously against Israel: the failure to win quickly and decisively
left Israel entirely isolated outside the United States.[23] Perhaps most
important, the moderate Arab regimes of Saudi Arabia, Jordan, and
Egypt—which had initially condemned Hizbollah's raid—soon
muted their criticisms of Hizbollah in the face of mass street
protests against the Israeli air campaign. Realizing after two weeks
that airpower alone was not working, the Israelis finally decided on
a ground invasion of southern Lebanon.

 The operational task before the Israeli ground forces was not
an easy one: essentially it mandated the reduction of a large number
of autonomous fortresses, a function of the terrain, tactics, and

organization of their enemy. Hizbollah's regular forces consisted of approximately one thousand full-time fighters,[24] many of them well trained by Iran and equipped with modern antitank missiles. These soldiers were Hizbollah's most mobile and centrally controlled forces. In addition to these elite forces, Hizbollah fielded an unknown but certainly larger number of local militias.[25] These forces were less well equipped and trained, and mostly operated out of the hilltop villages which dot the rocky, hilly countryside of southern Lebanon. In battle, these forces were largely immobile, confined to static defense of their particular villages, although there was some movement of these forces from the north into the theater before the Israeli ground invasion. In terms of numbers and quality, therefore, Hizbollah fielded the equivalent of an elite, over-strength infantry battalion supported by an unknown number of lower-quality local militias. However, any comparison of Hizbollah's forces to an organized military unit like a brigade or a battalion is potentially misleading, as Hizbollah's forces fought largely as autonomous rifle squads of seven to ten men and antitank squads of four to five.[26] While multiple Hizbollah squads coordinated in particular engagements, they were not organized as such into permanent "platoons" and "companies." Hizbollah had no pretenses of concentrating large maneuver forces to actively defeat the IDF; its goal was simply to survive and bleed the Israelis as much as possible while continuing to rain rockets on northern Israel.[27] To that end, it scattered the short-range surface-to-surface rockets and largely autonomous squads in manmade bunkers and the natural fortresses of the hilltop villages throughout southern Lebanon, where the Israelis would have to dig them out.

Hizbollah was thus a profoundly different enemy from both the conventional-force Arab armies the Israeli Defense Forces had historically confronted and the urban insurgency they face in Palestinian areas. Hizbollah was a decentralized force that depended on layers of static defenses in constricted terrain rather than maneuver. Its defensive firepower—buttressed by modern anti-armor weapons—was limited, but the overall effect was to diminish the advantages of Israel's most lethal, mobile, and usually effective units. Hizbollah and the Lebanese militias also demonstrated, and indeed relied upon, a

level of unit cohesion (though probably with deeply different moti-
vations) unknown in the Arab armies of past wars. The Israeli ground
offensive, moreover, slow to begin and slower to make progress,
failed to achieve the decisive military success required for Israel's
stated war aims. Hizbollah continued to hit northern Israel with rockets, including 250 on August 13, 2006, the last full day before UN
Resolution 1701 ended hostilities.[28] While the IDF did degrade the
Hizbollah forces and defenses in southern Lebanon, these losses can
certainly be made good. Perhaps most important, the longstanding
"moral superiority"—a combination of tactical, technological, and
psychological advantages—enjoyed by the IDF has been eroded, and
in particular its supremacy in the art of combined-arms conventional
battle has been called into question.

The IDF's troubles tell a cautionary tale that American ground
forces would do well to heed. To begin with, the Israelis seriously
misjudged the quality of their opponents, their tenacity, and the
extent of their defenses in southern Lebanon. Not merely mistaken
in the belief that an air campaign could achieve their objectives, the
IDF committed land forces to a battle for which many were poorly
prepared. The maneuver skills that were once the hallmark of the
Israeli Army had atrophied.

Consider the IDF's difficulties in concentrating its land attacks.
While well prepared and arrayed in some depth, the Hizbollah
defenses were essentially static. Hizbollah did not have any significant maneuver forces to complement its string of fortresses—a
weakness which the Israelis should have capitalized on. The
Israelis, with their superior numbers and mobility, should have
been able to concentrate their forces and attack each fortress or village with utterly irresistible local superiority—no Hizbollah fortress
or village should have been able to resist for long when the Israelis
decided to capture it. But Hizbollah held its fortresses, especially in
the beginning phases of the IDF ground invasion.[29] The Israelis frequently made small, local attacks on Hizbollah fortresses and villages that proved unequal to the task and subsequently had to be
"rescued" haphazardly by other units.[30] When the Israelis concentrated larger forces against their targets, as they did in the last few

days of the war, they had far more success at reducing Hizbollah fortified hilltop villages.[31]

Perhaps the biggest battlefield surprise of the war, however, came from Hizbollah's ability to limit the effectiveness of the Israeli armor. Israeli tank units had long been the pride of the IDF and had historically constituted an advantage for the Israelis. Hizbollah destroyed or disabled the IDF's heavily armored Merkava 4 tanks by effectively employing advanced antitank guided missiles (ATGMs). According to the Israelis, fifty-two Merkavas were hit, twenty-two were penetrated, and five of those were destroyed.[32] Those are significant operational losses.[33]

In his insightful monograph, *Hizballah at War: A Military Assessment*, Andrew Exum states that the 2006 Lebanon War "will forever be the war of the antitank missile," a common assessment.[34] But this is a summary judgment—its clever brevity masks a far more worrisome set of factors. In fact, the Israelis' problems were not merely technological. The Hizbollah missile teams were well trained: generally, Hizbollah damaged the tanks' tracks first, and then methodically "executed" them by firing again and again at their weak spots—the destruction of Israel's Merkavas resembled the laborious sinking of the crippled *Bismarck* more than the quick destruction of the HMS *Hood*. The problem was not that Israeli tanks were too vulnerable to individual hits, but rather that the Israelis allowed Hizbollah to hit the same tanks over and over again.

Most assessments of the 2006 Lebanon War have rightly criticized the operational employment of Israel's armor. Israeli commanders, it is charged, failed to recognize that southern Lebanon is not conducive to maneuver warfare and instead expected a repeat of Israel's lightning victories in past conventional wars against Arab armies. In those wars, Israel achieved great success with sweeping armored thrusts, as they were fighting in open tank country, and the quality of the enemy infantry was low. Furthermore, the conventional armies of their enemies were susceptible to destruction by envelopment and were dependent on lines of communication. In the 2006 Lebanon War, Israel's tanks were far less effective in that role, as they faced a more resolute and tactically sophisticated enemy with better infantry antitank weapons in more difficult terrain. The rocky, hilly battlefield restricted the Israeli armored forces to predictable avenues of advance.[35] Hizbollah's antitank teams were able to ambush Israel's tank columns as they snaked down the roads and *wadis*. Furthermore, Hizbollah was able to mine the few main routes of advance with thousand-pound antitank mines. Even if the roads and *wadis* had not been so dangerous, most of Hizbollah's forces were holed up with ample supplies in fortifications and hilltop villages—"breaking through" to the enemy's rear, the traditional role of armored formations, is of limited value when the enemy consists of a number of autonomous fortified points which do not require constant resupply via lines of communication. There were simply few opportunities to bypass Hizbollah defenses and collapse them from the rear.

Thus, as most assessments of the 2006 Lebanon War point out, the IDF armor by itself was not enough—and indeed proved itself vulnerable—without well-integrated and dismounted infantry support.[36] Hizbollah and the Lebanese militias cleverly combined their antitank missile fires with their infantry strongpoints; by contrast, the IDF often employed its armor without adequate infantry support. Thus the pace of the fight was slow: just what Hizbollah wanted and just what the Israelis did not. In a sense, it was a World War I tactical situation that involved fighting for every yard; but with twenty-first century technologies, initial engagements could occur at very long ranges indeed. With a shortage of dismounted infantry, the Israelis had difficulty stripping out the Hizbollah forward defenses, with the result that its armor lost its mobility and became vulnerable.

But if the Israelis' problems went beyond those of technology and the challenges posed by advanced ATGMs, they also went beyond the tactical performance of Hizbollah: this war exposed institutional problems that confront the IDF—and may also confront U.S. land forces. Simply put, and with important exceptions in several units, the IDF was not well prepared for large-scale, high-intensity combat operations. This lack of preparedness was demonstrated at two levels: that of the Israeli military as a whole, and that of the ground forces in particular.

Like the U.S. military of the post–Cold War period, the IDF has developed an aversion to extended land warfare. Issues like sensitivity to casualties are even greater in Israel than in the United States, while at the other end of the spectrum, the growing Iranian nuclear and missile arsenals are a daily and ever-present threat: the IDF has a lot on its plate. And then there is the unending irregular conflict with the Palestinians, made worse by the rise of Hamas.

Thus, in the six years preceding the conflict, the Israeli military severely degraded its ability to fight a high-intensity ground war. In 2000, the Israelis ended their long occupation of southern Lebanon. A cult of airpower emerged, promising that the Israeli Air Force (IAF) could prevent the advent of high-intensity ground warfare by acting as a deterrent. Furthermore, the proponents of airpower believed that should the IAF fail in its deterrent role, it could still eliminate the need for high-intensity ground warfare by decisively defeating the enemy on its own, save for a few special forces to aid in finding and tracking enemy assets.[37] Consequently, from 2000 to 2006, the IDF transferred resources from Israel's ground forces to the IAF.

But the Israeli land forces also bear much of the responsibility for diminished readiness. Israel's withdrawal from Lebanon in 2000 coincided with the beginning of the Palestinian *intifada*. The Israeli ground forces became so focused on policing the Palestinian uprising that they failed to prepare for the possibility that they might once again find themselves in high-intensity ground warfare. Israeli tank crews, for example, were often dismantled and used as light infantry in the occupied Palestinian territories rather than training with their vehicles. During the 2006 Lebanon War, many tank commanders would complain that their tank crews (especially the reservists) were not adequately trained in their designated roles as tankers. Likewise, even though Israel's infantry demonstrated excellent élan in the 2006 Lebanon War,[38] their training and experience with the Palestinian irregulars did not prepare them for larger and more sustained combat operations, such as the storming of Hizbollah's fortified hilltop villages.[39]

American ground forces today find themselves in a similar dilemma: not only are they stretched nearly to their limits by the demands of two irregular conflicts in Iraq and Afghanistan, but they must retain their advantages in large-scale maneuver warfare and prepare for even more sophisticated operations in the shadow of nuclear weaponry and ballistic missiles. Moreover, at the level of pure tactics, it is a certainty that future enemies of the United States will attempt to emulate Hizbollah.[40]

Hizbollah represents a dangerous cross between a conventional army and an irregular insurgency. Unlike a conventional army, it is, if not immune, then very resistant to airpower.[41] Unlike old Arab armies, it is willing and able to stand and fight against a conventional army on the ground. For the task of prying such a determined foe out of defended positions in difficult country, infantry is the most important combat arm. Armored formations are necessary but not sufficient; autonomous combat forces like Hizbollah do not require constant resupply via traditional lines of communication, a fact which negates the basic advantage of armored forces at the operational level—their ability to neutralize enemy forces through maneuver and envelopment. At the tactical level, Hizbollah's employment of more sophisticated weaponry and better leadership and unit cohesion erode historical Western advantages. The United States must understand and be prepared to meet the challenges such a force poses. As U.S. land forces adapt to their own serial irregular wars, they must remember to retain the ability to fight in a high-intensity, combined-arms campaign.

Lost and Won: The Fight for Anbar

In February 2006, a massive bomb was detonated outside the Ali Askari mosque in the Iraqi town of Samarra, sparking a period of sectarian violence centered in Baghdad but raging throughout central Iraq. The conflict, which prompted many to declare a "civil war,"[42] paralyzed the Iraqi government and made even tentative efforts at political reconciliation impossible. In Washington, a political establishment already soured on the war all but declared defeat. The Baghdad situation, however, paled in comparison to the chaos just west of the capital, in the Sunni hinterlands: in September

2006, a marine official told the *Washington Post* that "the prospects
for securing that country's western Anbar province are dim . . . there
is almost nothing the U.S. military can do to improve the political
and social situation there."[43] Yet by spring of 2007, the situation
had been reversed entirely:

> Ramadi was the most dangerous city in Iraq . . . Now, a
> pact between local tribal sheiks and U.S. commanders has
> sent thousands of young Iraqis from Anbar Province into
> the fight against extremists linked to Al Qaeda in
> Mesopotamia. The deal has all but ended the fighting in
> Ramadi and recast the city as a symbol of hope that the
> tide of the war may yet be reversed to favor the Americans
> and their Iraqi allies.[44]

How did this happen?

The turnaround of 2007 was a product of changed Iraqi attitudes
and combined military operations by the Iraqi army and U.S. Army
and Marine Corps units. But it is the Marine Corps that has seen
the extreme best and worst of times in the long-running struggle
for Anbar. Since the invasion's initial success, marines found themselves in a strategic stalemate against Sunni insurgents and
Al Qaeda in Iraq in the province. The marines were relatively few,
and Anbar is an expansive battlespace: Anbar accounts for 30 percent of Iraq's landmass, borders Jordan and Syria, and contains
Ramadi and Fallujah, two key economic and military population
centers along the Euphrates River. The population is predominantly
Sunni, and Anbaris generated the initial momentum for the Sunni
insurgency after the American invasion. Previous attempts to pacify
the region with large assault teams failed to translate the American
tactical superiority into persistent success against the insurgents'
modus operandi: mobile logistics centers, weapons caches, improvised explosive devices (IEDs), and light weapons raids. For two
years, efforts to tamp the violence in Anbar failed utterly as AQI
operated freely from safe houses and weapons centers to infiltrate
the tenuous local governments. While American military attention

has centered on Baghdad, Anbar has remained the most dangerous area for American soldiers, with 1,279 confirmed deaths.[45]

The marine campaign to retake Anbar began in 2006, when marines supported by the army's First Armored Division began major operations along the Euphrates River. Fighting was relatively light, and the American tactical advantage (including coordination between airpower and ground forces) prevented insurgents from mustering effective counterattacks. As insurgents "went underground" the marines began an occupation phase along the newly captured cities. Almost immediately after the major operations ended in Ramadi, vehicles traveling along main roads began sustaining casualties from a mix of IEDs and insurgent ambushes. American commanders anticipated lightly armed guerrilla units, so the growing use of IEDs, the newer and more lethal explosively formed penetrator mines, and coordinated ambushes surprised the Humvee-dependent marine forces. Again, numbers mattered: although marine commanders knew the estimated enemy strength, they lacked the personnel and equipment necessary to adapt to the ever-changing urban environments and enemy weapons capabilities.

The fighting in Iraq has tested every potential capability of the deployed armed forces. Marine units, in particular, have been called upon to perform varying mission types, on varying terrain, with varying intelligence, and with varying equipment. A unit might root out suspected insurgents and arms dealers one week, and then offer them weapons training the next week to provide security. A general trend has emerged, however: versatile, motivated, and innovative officers are essential at every rank to cope with a variety of unexpected contingencies. Although technology and manpower account for American military success, a leader's training and creativity truly set his unit apart.

Al Qaeda in Iraq had long possessed large swaths of rural Anbar, but by 2006 its influence had begun to seep into the dilapidated cities with defunct municipal authorities. These towns, ramshackle to begin with, had by now seen several years of intense fighting and were essentially ungoverned spaces: they had no persistent U.S. presence and few Iraqis. Ramadi, a city of three hundred thousand, lacked a mayor and provincial government. Following the first assault on Fallujah in 2004, the number of insurgent in the area skyrocketed, and al Qaeda easily established bases along the unoccupied banks of the Euphrates River.

As the violence climbed through the summer of 2006, U.S. commanders realized that the key to regaining Anbar lay in separating the al Qaeda extremists—whose first priority was to continue the war—from the traditional Anbari tribal authorities—whose priorities were to defend their people and their own power. Marine unfamiliarity with the terrain and populace (marines had been deployed to Anbar for years, but unit rotations of six months crippled efforts at continuity) led army colonel Sean MacFarland of the First Brigade, First Armored Division, to search for allies among the thirty-one Sunni tribes scattered throughout Anbar. American attempts to reach out to Sunni sheiks in the two previous years had been unsuccessful. American commanders had held a "breakthrough" conference with local leaders, but no definite alliances were formed or reliable intelligence for enemy identification gained. Battling AQI and local tribes while simultaneously attempting to improve municipal services bred pessimism among marine commanders, who had entered the fight expecting to be soldiers, not diplomats.

But in fact, there was a nascent opportunity for local diplomacy and alliance building: al Qaeda had begun to overplay its hand. In January 2006, a string of AQI attacks fostered resentment among local leaders. A suicide bomber killed seventy locals at a police recruiting drive; insurgents killed four Sunni sheiks for cooperating with the Americans; after assassinating one prominent Sunni sheik, they blatantly offended the community by refusing to allow the family to bury his body for four days. A group of tribal leaders led by Sheik Abdel Sittar, announcing himself a representative of the Anbar Awakening council, approached Col. MacFarland to offer support against al Qaeda. The sheiks presented a twelve-point plan outlining their expectations of the Americans, including guarantees of political representation and financial aid for municipal projects. Sittar promised MacFarland that if the stipulations were met, the community would embrace the Americans. Sittar immediately mobilized forty-five hundred local tribesmen to join the paltry Ramadi police force, and—possibly of greater value—a steady stream of intelligence now flowed to the Americans. Col. MacFarland described the Anbar Awakening as "a grass-roots, democratic response to al-Qaeda . . . Shortly thereafter, caches started popping up like weeds, essentially disarming these tribal areas."[46]

In Anbar, reliable intelligence had been nearly impossible to obtain. In mixed ethnic areas, U.S. commanders could exploit militias by using sectarian tensions to gain actionable intelligence on rival factions. In a homogenous Sunni region under the control of AQI, there were no such "seams" to exploit. Once the sheiks began offering their assistance, however, a steady stream of intelligence flowed to the marines. The sheiks' knowledge of local alliances proved invaluable in driving a wedge between AQI and the local populace. American commanders almost exclusively depended on local sheiks to identify al Qaeda fighters, and a more concerted marine focus on reconstruction and security forces training followed.

An influx of U.S. units reinforced these successes. When the majority of the First Armored Division's First Brigade Combat Team entered Ramadi in 2006, Col. Sean MacFarland immediately sought to establish a greater force presence by building small combat outposts in the city's worst neighborhoods. The eighteen combat outposts allowed the unit to secure Ramadi "a chunk at a time."[47] Still, the first price of victory was measured in increased fighting and casualties. The marines suffered daily attacks at the lightly fortified, isolated, joint-security stations. MacFarland placed a battalion under Lt. Col. V. J. Tedesco in the southern part of the city, and it lost twenty-five tanks, Bradley fighting vehicles, and trucks due to IEDs. The brigade entered operations with fifty-five hundred soldiers and lost ninety-five, with over six hundred wounded during its tour.

Despite these costs, the progress was rapid and real. In March 2007, Iraqi Prime Minister Nouri al Maliki visited Ramadi and met with Sittar to discuss future Sunni representation within the national government. By mid-March, U.S. forces reported that they had pushed insurgents farther east and north of Ramadi's center.

The number of police recruits in Ramadi jumped from about thirty a month to one hundred in June 2006 and about three hundred in July. More than three thousand new recruits had joined the police by the time MacFarland's brigade left in February 2007,[48] and marine Brig. Gen. Charles M. Gurganus reported that by May 14, 2007, at least twenty-two joint security stations existed within the city.[49] U.S. forces and Iraqi police, acting on military intelligence and local tips, were able to reduce violence by 70 percent.[50] In March 2007, marine Maj. Gen. Walt Gaskin stated: "Our strategy of clear, hold, and build, combined with an energized government and tribal engagement, is beginning to bear fruit."[51]

To be sure, the future in Anbar is uncertain. Despite political progress, the long-term security situation remains a continuing concern. The insurgency relies on economic and personal insecurity to foster sectarian resentment and attract new recruits, and only deliberate efforts to maintain working relationships with the sheiks have enabled the American military to keep the insurgency from regaining ground. U.S. commanders realize that building long-term, stable, and self-supporting local municipalities is possible only if all areas are kept from the insurgents.

While long-term improvement in Anbar rests with the Anbaris, there are several lessons to be extracted from the progress achieved thus far. The first is this: while in counterinsurgency operations, tactical military success is no guarantee of a lasting victory, security must be established before political progress can be cemented. In practical terms, "protecting the populace" means securing the places where people live and work. It takes time, it takes manpower, and it means driving out the enemy. Units designed to be expeditionary—to deploy and redeploy rapidly—are undeniably handicapped by the demands of long-duration operations. The Marine Corps policy of six-month rotations in Iraq, determined by the longstanding operational concepts and personnel policies of the corps, makes it much harder to sustain the holding operations that now have proven so fruitful in Anbar. It also makes building the local political alliances, such as the partnership with the Anbar Awakening movement, that much more difficult. At the same time, there is a symbiotic quality to the alliance that may hold it together over time: Sheik Sittar claimed that the marine presence lent the Anbar effort credibility in the eyes of the national government, while the sheiks' embrace gave marine units legitimacy among the local population. But the September 2007 assassination of Sheikh Sittar shows how difficult it can be to achieve a self-sustaining and decisive result.

Second, the Marine campaign to retake Anbar province illustrates the necessity of flexibility at the tactical level and the value of initiative among junior leaders. The conflict in Iraq emphasizes the human dimensions of war far beyond the technological. At the same time, it does not matter what a weapon, unit, or tactic was designed to do, but rather what it can do. While conventional operations have proved essential, such as in Fallujah in 2004, the marines' ability to adapt to the conditions of irregular warfare has begun at the lowest levels. Officers and NCOs at the company level and below—historically a strength of the Marine Corps—are the most crucial leaders. A corollary is that the corps needs to rethink its reliance on one-term enlistments; a corporal may have strategic effects, but unless he gains greater maturity and experience, too many of those effects may be unintended and problematic.

Finally, the experience in Anbar—and indeed throughout Iraq and Afghanistan—suggests that the traditional American idea of intelligence collection is not well suited to small wars. The reliance on technical means, the effort to centralize analysis, the desire for a complete understanding of the battlespace, and, most of all, the separation of intelligence from operations have too often left American units lost in the various fogs of war. The marines found in Anbar—as units have too infrequently and too late discovered elsewhere—that presence and the holding of ground are the key to better and more "actionable" intelligence. Units must act to create intelligence, not gather intelligence first and then act. In a conflict where there are very few large enemy formations and the tactical situation is so fluid, knowledge is evanescent, fleeting, and invariably incomplete.

Building Partners: The Abu Sayyaf Campaign

Since early 2002, the U.S. military has assisted the Philippines in its struggle with the Abu Sayyaf Group (ASG), an al Qaeda affiliate active in the southern Philippines, through Operation Enduring Freedom–Philippines (OEF-P). OEF-P can be understood as a manifestation of the long—and often troubled—relationship between the United States and the Philippines. It exemplifies the opportunities and challenges of reframing or creating new alliances in the Long War; a lasting victory may well depend on the ability of U.S. land forces to create partnerships with militaries like the Philippine Army. The indirect approach, working with and through others, may be more important than direct combat conducted by U.S. forces, since the fight against global terrorist networks demands a level of persistence, presence, and local knowledge best provided by local troops.

The Abu Sayyaf campaign highlights the duration of these local conflicts. The group was founded in 1991 by Ustadz Abdurajak Janjalani, one of several hundred Moro[52] fundamentalists who served with the *mujahideen* in the Soviet-Afghan War. Janjalani has been linked to the forty-eight-person Executive Council of the Islamic International Brigade, the forerunner to al Qaeda.[53] Following the end of war in 1989, he was encouraged by Osama bin Laden to return home, wage *jihad*, and create a *sharia*-based Islamic state in the southern Philippines. In August 1991, the newly formed Abu Sayyaf launched its first attack, a grenade strike on Christian missionaries. This assault proved to be the modus operandi the organization would follow through 1995. For four years Abu Sayyaf waged a fearsome sectarian terror campaign against the Philippine Christian community, bombing numerous churches and kidnapping clergy, missionaries, and parishioners. The reign of terror killed 136 people and wounded hundreds more.[54] Throughout this time al Qaeda supplied Abu Sayyaf with funds via the Islamic International Relief Organization (of which Jamal Khalifa, Osama bin Laden's brother-in-law, was the regional director).[55]

By the mid-1990s, however, Abu Sayyaf had to abandon its traditional tactics. The capture of Ramzi Yousef, their principal al Qaeda emissary, deprived the organization of financial support and forced it to adopt, simply to survive, a program of kidnapping for ransom. The group's *jihadist* mission became even more tenuous with the death of founder and chief ideologue Abdurajak Janjalani in 1998. For all intents and purposes, ASG had abandoned *jihad* in favor of mere banditry. One Abu Sayyaf defector said of the period: "The group lost its original reason for being. The activities were not for Islam but for personal gratification. We abducted people not . . . for the cause of Islam, but for money."[56] Nevertheless the organization survived, even though the main proponents of the kidnap-for-ransom method within Abu Sayyaf had been killed by January 2003. Their deaths paved the way for Khadaffy Janjalani—the younger brother of the group's founder—to assume leadership of the group. Under his command, Abu Sayyaf returned to its sectarian terror roots, regained funding from al Qaeda, and resumed politically and religiously motivated terror attacks, spectacularly culminating in the SuperFerry bombing in February 2004 that killed 194.

The U.S. military began working with the Philippines Armed Forces in its struggle with Abu Sayyaf in early 2002.[57] Yet Philippine history complicated the arrangement; the former occupation of the Philippines by the United States, and the presence of massive military bases there even after independence, have made the government highly protective of its sovereignty and imposed constraints on the level of American assistance. The Philippine constitution prohibits combat operations by foreign forces on homeland soil, and as a result, the United States has been able to combat Abu Sayyaf only indirectly, by supporting Philippine forces. Furthermore, many Filipinos feared that the United States would use active participation in this campaign to justify the reestablishment of permanent military bases in their country.[58] Unsurprisingly, when the United States proposed in 2003 that its forces "conduct or support combat patrols," the idea was met with uproar in the Philippines and abandoned. Thus, Philippine support for operations is conditioned on a

light U.S. footprint, and military efforts have been limited to training, advising, and equipping the Philippines Armed Forces, as well
as undertaking reconstruction projects and gathering intelligence.

Despite these political restrictions, the Philippine edition of
Operation Enduring Freedom was initially tremendously successful. In 2002, an effort dubbed Operation Balikatan 02-1 built upon
traditional training missions to clear Basilan, the northernmost
island of the Sulu archipelago, an island chain stretching southwest
from the Philippines towards Malaysia and Indonesia. The U.S.
forces totaled approximately thirteen hundred personnel and were
key to making a successful Philippine offensive possible. Upon
arrival on Basilan, U.S. forces found the Philippine units on that
island in disarray, unable and unwilling to aggressively pursue Abu
Sayyaf. The first step was to deploy Special Forces to gather intelligence as unobtrusively as possible. In traditional Green Beret
fashion, they did not confine this effort to locating Abu Sayyaf
assets and personnel, but also collected extensive demographic and
economic information on the island's many villages. The second
step was to improve the capabilities of Philippine forces. American
advisors helped their Filipino counterparts make bases more defen-
sible and worked with Philippine soldiers and marines on more
advanced aspects of fieldcraft, such as combat lifesaving skills.[59]
This training had a pronounced and positive effect on Philippine
morale, paving the way for the crucial step: securing the populace
and separating them from the insurgents.[60]

Philippine forces began to patrol more aggressively, establish
security at the village level, and clear Abu Sayyaf safe havens.[61]
Consistent with classic counterinsurgency practice, this clear-and-
hold phase was followed by efforts at reconstruction. Using the previously gathered village demographic and economic intelligence,
Special Forces and civil affairs soldiers conducted small-scale
humanitarian and civic action programs that yielded immediate
improvements in the targeted villages' quality of life. The efforts to
improve the local inhabitants' lives earned their trust and led to an
increase in intelligence tips. As the security situation improved, U.S.
commanders deployed a U.S. Naval Construction Task Force to

execute larger-scale reconstruction projects such as road and bridge building.[62] These bigger projects, in turn, further endeared the national government to the villagers—U.S. units took great care to help establish the legitimacy and boost the competence of the Philippine forces and government—and thus began to drive a wedge between Abu Sayyaf and the local populace. Abu Sayyaf's strength dwindled from an estimated twelve hundred combatants at the beginning of the year to only two hundred to four hundred by year's end. Ultimately the terror group was forced to abandon Basilan—formerly its major stronghold—and relocate to Mindanao.

Yet the very success of operations on Basilan in 2002 also underscores the long-term challenges of partner building. The battered Abu Sayyaf has relocated to what amounts to an unassailable refuge: territory controlled by the Moro Islamic Liberation Front (MILF) on Mindanao island. While Philippine forces continue to attack Abu Sayyaf whenever they find them, they lack the capacity to launch a comprehensive clearing campaign against the group's sanctuary on Mindanao or to replicate their successful Basilan campaign. And although Abu Sayyaf's armed strength has hovered at two hundred to four hundred combatants since 2002, it has increased its campaign of terror bombing, focusing on civilian urban targets. Equally troubling is the fact that Abu Sayyaf's bombmaking and operational techniques have improved considerably.[63] The Abu Sayyaf Group has proved to be extremely resilient and more than capable of rapidly replacing its poorly trained and expendable low-level operatives.[64] Furthermore, the quality and integrity of the Philippines Armed Forces remain a concern. Some units fight exceptionally well, while others leave much to be desired. Cornered Abu Sayyaf forces have "mysteriously" slipped through Philippine cordons on a number of occasions, and have bribed their way out of government detention centers.[65]

The principal reason for Abu Sayyaf's survival, however, has less to do with its capabilities than with the fact that it has become intertwined with the Moro Islamic Liberation Front and another larger insurgent group, the National People's Army (NPA). The MILF association is by far the most important for Abu Sayyaf. The Moro peoples have long resisted outside government; they did so during the Spanish and American colonial periods, and have continued their resistance since Philippine independence. Founded in 1978 as a radical offshoot of the Moro National Liberation Front,[66] MILF desires to carve out an independent state ruled by Islamic law in the southern Philippines. While Abu Sayyaf and MILF share the same goal, they differ substantially in their methods. MILF is a classic rural insurgency, using methods that recall Mao Zedong's "people's war." It employs guerrilla warfare to deny or reduce the government's influence in areas populated by Muslims while at the same time attempting to bolster its own legitimacy by operating a shadow government and providing services in those areas. Moreover, its desire to win the hearts and minds of the locals has led it to disavow terrorism and sectarian violence and profess to fight a "clean" guerrilla war against the government. And its local military power is formidable: it can mobilize a force of roughly twelve thousand fighters.

Yet despite these major operational differences, MILF has provided refuge to Abu Sayyaf's leaders on Mindanao since 2002, and has even allowed the terror group to recruit new *jihadis* on its territory; this patronage makes it extremely difficult to finish off Abu Sayyaf. The Philippine government is reluctant to strike at the operatives hiding in remote bases in MILF-controlled territory. And indeed the Filipinos' own strategy for dealing with the Moros almost rules out a more vigorous anti–Abu Sayyaf campaign: the central government would prefer to come to a political accommodation with the MILF. After decades of conflict that killed tens of thousands, the government and MILF entered into peace talks in 2001. Neither side has wholly respected the cease-fire, but the level of violence has declined by several orders of magnitude. The Philippine government is reluctant to undermine those talks by attacking MILF's territory. Dealing with Abu Sayyaf's few hundred terrorists simply is not worth a provocation that would ruin the government's chances of reaching a permanent settlement with MILF, a much more formidable opponent.

Abu Sayyaf is also an indirect beneficiary of the New People's Army, the armed wing of the Communist Party of the Philippines, which is waging a Maoist "people's war" against the government in the northern hinterlands of Luzon.[67] Like MILF, the NPA is a conventional insurgency that controls territory, operates a shadow government, and fields upwards of ten thousand guerrilla fighters. Since the end of the Cold War, the United States has stopped viewing Maoist insurgencies as a major security issue, but the government of the Philippines believes that combating the NPA is its most urgent security priority.[68] The ongoing struggle with the NPA further constrains the ability of the Philippine Army to deal with MILF and Abu Sayyaf.

Despite the tremendous tactical and operational successes achieved in Operation Enduring Freedom–Philippines, the current situation reveals the challenges inherent in achieving a larger effect. First, American objectives cannot override local imperatives; the political basis for building strategic partnerships must naturally be dominated by the interests of the partner government. This does not mean that common objectives are impossible, or that American interests will not be taken into account by our partners, but simply that the United States cannot call all the shots. Our purpose must be to buttress the legitimacy of the partner government and military. Second, it is clear that lasting effects can be achieved only over a long course of time: the key factor is convincing both our allies and our adversaries that our commitment to their country will indeed be enduring. Finally, the experience of beating back Abu Sayyaf in the Philippines demonstrates the complexity of trying to combat terrorists through another government's armed forces.

To consider this last point more closely: Operation Balikatan 02-1 in 2002 was a great success, clearing Abu Sayyaf from its major stronghold on Basilan and reducing its power. The Special Forces assessment teams dispatched by U.S. Pacific Command in October 2001 did an excellent job gathering intelligence. U.S. advisors quickly trained the Philippine forces on the island into a more effective fighting force with improved morale, which enabled them to take the initiative from Abu Sayyaf and patrol aggressively to establish security at the village level. Improved security allowed U.S. forces to begin targeted reconstruction projects that quickly helped turn the islanders' hearts and minds against Abu Sayyaf. This shift in the populace's attitude doomed the group's position on the island.

But after these initial victories, the challenges have become much harder—that is, much harder in the view of the Philippine government. That government shares the United States' desire to deal with Abu Sayyaf, but with Abu Sayyaf now relocated to MILF-controlled territory on Mindanao, it is unwilling to do so; it does not wish to risk disrupting the MILF peace process, especially while it remains locked in a bitter struggle with the NPA, which the Filipinos regard as more threatening to the government than Abu Sayyaf. Thus, to achieve further success against Abu Sayyaf (and to forestall the prospect of strategic cooperation between Abu Sayyaf and the NPA), the United States needs to assist the Philippine government in solving its NPA problem. But this is exactly the definition of a genuine strategic partnership: assisting the indigenous government and military in its efforts to establish its legitimacy and extend its writ. To be effective, American forces must undertake longer-term efforts not simply to track down terrorist groups that immediately threaten our interests, but to help the Filipinos deal with the insurgencies they find most threatening. This means a continuing military-to-military investment: the Philippine Army simply lacks the resources to deal with these threats simultaneously.

Ironically, while the peace process initiated by the Philippine government with the MILF has inhibited military operations against Abu Sayyaf, it is ultimately the United States' best hope for eventually dealing with the terror group; as is so often the case in a successful counterinsurgency campaign, political and military efforts must reinforce one another. The goal is to detach Abu Sayyaf from MILF, and this will happen only if the government and MILF reach a political settlement. And the purpose of buttressing the Philippine Army's capacity is not to "conquer" the Moros—historically a fruitless enterprise—but to shape the political negotiations in a way that better integrates the Moros in the Philippine polity.

4

What Kind of Force?

Given the number and variety of missions confronting the United States and the emerging nature of land war, it is apparent that U.S. land forces need not only to be more numerous but also to possess quite different qualities than simply the timely and devastating delivery of firepower. If the Pentagon's transformation model was for rapid, decisive operations, our post-9/11 experience tells us there can be no one-battle war. The kinds of conflicts described in chapter 3 more resemble the frontier fighting of the nineteenth century—not only in the American West but in the far-flung outposts of the British Empire—than they do the epic clashes of European armies in the twentieth century. Indeed, the land forces of the United States face a quandary not dissimilar to the classic conundrum of the British Army, that is, how to find the right balance between the requirement for large-scale conventional operations, what the British called their "continental commitment," and the requirement for smaller-scale operations on a faraway security perimeter, what the British called their "imperial commitment." Today and in the foreseeable future, the pendulum is unquestionably swinging toward the small-wars pole. Such missions call for qualities that differ significantly from those given priority in past land forces.

At the heart of this new proposition may lie a profound change in the compact between U.S. land forces and the American people, and a shifting basis for an all-volunteer force. It has been often observed in recent years that, while the military is at war (or, more acidly, that the army and Marine Corps are at war), the nation is not. Indeed, except for the complications of airline travel or in the communities clustered around military installations, and for all the political energy invested in the debate over Iraq policy, the domestic effects of Iraq and Afghanistan are often hard to find. It is one thing to sustain and succor a professional military under the circumstances of the later Cold War, when the mission was to deter war; quite another to fight the Long War, when serious and significant numbers of casualties are a constant prospect. This has not been, and hardly can be, a perfectly shared sacrifice; the real links between soldiers and society at large must somehow be strengthened, even in circumstances that militate otherwise.

The worst solution is the one most discussed. A return to the draft would harm military effectiveness while jeopardizing the social compact. To send hastily trained, short-service conscripts to patrol the streets of Baghdad or the hills of Helmand province would simply put more Americans in harm's way without real purpose or value. At the same time, we must understand that the missions of the Long War—constant, long deployments in life-threatening environments—are placing new strains on people in uniform. Moreover, we Americans must do more than simply admire our soldiers and marines, as from afar: more of us must join them, serve alongside them. This sense of obligation ought most to be shared among the American elites, but should be shared across society as a whole. At minimum, we should provide sufficient resources to the military services; budgets are expressions of not only a nation's fiscal priorities but its moral values. When the distance between the frontier and the home front is so great, it falls primarily upon those at home to bridge the gap.

It may now be necessary to revisit some of the basic tenets of the all-volunteer force (AVF). Is it reasonable or even possible to offer soldiers and marines the same simulacrum of American middle-class life that was the unwritten standard of the late Cold War years? In the late 1980s, a tour in Korea was a one-year, "hardship" tour; compared to the constant rotations to Iraq and Afghanistan, such a tour seems like a relative luxury. During the AVF years, the force has not only grown older—and a more mature perspective is certainly of value in irregular warfare—but is more likely to be married and to have larger families. Back in 1993, former Marine Commandant

Gen. Carl Mundy was roundly criticized for attempting to formalize the service's preference for single people, particularly in the case of young marines; he was forced to recant and apologize to the secretary of defense, Les Aspin, who according to his spokesperson wanted to "balance readiness and family issues."[1] This balance has been a key to the AVF, producing a force that has proven remarkably durable and successful, with superb units in both the army and Marine Corps. Tampering with it should not be lightly done; at the same time, the goal is to build the force to fight the war we have, not select the war that suits the force.

How precisely to strengthen the relationship between soldiers and society at large, and what precisely the all-volunteer force should look like, are questions that go beyond the boundaries of this study. But as we consider the qualities and quantities needed to maintain sufficient American land power, we would do well to keep in mind the first-order issue of the role of military forces, and especially land forces, in our society.

Force Presence and the Institutional Base

Today's conflicts place a premium on units that are robust enough to operate independently and with initiative, at the extremes of the American security perimeter. Because the dominant mission for U.S. land forces is to win the Long War—often described as a "global counterinsurgency"—the army, Marine Corps, and Special Operations Forces must reflect the characteristics of successful counterinsurgency forces. These characteristics turn the transformation model almost completely on its head: for land forces especially, continuous presence is a higher virtue than the ability to strike rapidly from great distances. As David Galula wrote in the 1964 classic *Counterinsurgency Warfare: Theory and Practice*, forces that are a constant presence are preferable to those that are highly mobile. "The static units are obviously those that know best the local population, the local problems . . . It follows that when a mobile unit is sent to operate temporarily in an area, it must come under the territorial command."[2] What is true at the tactical level is also true strategically:

presence is a precondition to success. And what is prescribed by theory has been borne out by hard experience in Iraq and Afghanistan. If there is a single lesson about progress in those irregular wars, it is this: good things happen most often when U.S. forces are there, and bad things happen most often when they are not there. Insurgents do not seek out direct confrontations with U.S. forces; they avoid them, except in ambushes. Iraqi forces are exponentially more effective in partnership with American forces, and even the gung-ho Afghan National Army relies heavily on U.S. help. To again draw a contrast to the Pentagon's past "transformational" ideal of deploy-fight-recover, where U.S. forces sally forth from and quickly return to their home bases, the future will see not only forward-deployed forces but forward-stationed forces. To win, we must be there.

There are a number of methods for achieving presence, but the most effective and efficient method is to have forces that are stationed forward. This is simple common sense, and is also confirmed by experience: during the Cold War, it was far easier to defend the West German border with units and forces stationed in Germany. In the course of the Long War, we will achieve greater success at a lower cost as we are able to create a similar infrastructure along or near the American security perimeter; it is time to revisit recent assumptions about the need for in-theater bases. Increasingly, the model of presence has been one of forces based in the United States and deployed for relatively short periods abroad; it is as if the traditional marine or even airborne expeditionary model has been adopted across both forces. For the army, this contrasts significantly with the general experience of the late Cold War, when a substantial portion of the force was garrisoned in Europe in tours lasting two years or longer; even the "hardship" unaccompanied tour in Korea was a year. Until the need for surge forces in 2007 lengthened army tours in Iraq and Afghanistan to fifteen months, rotations had been capped at one year. Marine tours remain limited to seven months, compared to the traditional pattern of six-month "pumps," even though these relatively short deployments limit the effectiveness of marine units in the field. But such a churning "personnel tempo" is a very costly method of sustaining presence. The rotation ratio of total field force to deployed force is at least three to one; for every soldier or marine actually fighting in Iraq or Afghanistan, one is preparing to deploy while another is returning from deployment. And the ratio of the total active force—that is, the field force plus the "headquarters" force—to deployed force is approximately five to one, and even higher when levels of reserve component mobilization rise.

There are other models of force generation, however. Britain's Cardwell system, so-called for Edward, Lord Cardwell, secretary of state for war between 1868 and 1874, was a series of reforms driven in good measure by the need to feed regiments employed on constabulary missions throughout the empire. While a direct analogy to what America requires would be overdrawn, the British system expanded the size of units while slowing the pace of unit rotations, making more efficient use of limited manpower resources. The British Army also stationed its main regimental headquarters in the theaters where they were to operate. Alas, the post–Cold War American trend, accelerated by the transformational ideal, has been to consolidate the U.S. military at home. Thus, we find ourselves driven by the need to generate, deploy, and sustain forces at great distance; while generally not forward based, our land forces are required to conduct forward operations. We must reestablish a network of forward bases along the new American security perimeter, and we must begin to reverse the consequence of recent shortsighted decisions about troop presence and deployment.

The ability to generate and sustain forward presence and forward operations thus depends upon the capacity of the army and Marine Corps as institutions. Throughout the post–Cold War period—and even before—the institutions of the services have been under attack. The 1986 Goldwater-Nichols legislation diminished the importance of the Joint Chiefs of Staff and, more importantly, of the service chiefs of staff, not only by reducing their effect on operations in the field but also by strengthening the planning, programming, and budgeting authorities of the theater combatant commanders. Moreover, the combination of budget reductions and the mania for organizational efficiency—quite distinct from organizational effectiveness—mandated constant reductions in the institutional overhead of the services. For example, U.S. Army Training and Doctrine Command—which not only trains the entire army but is the key to developing the organizing and operating principles of the service—has been a constant target of the base realignment and closure process and the outsourcing impulse. Indeed, contractors are playing an increasing role in the production of army doctrine, the formal expression of service thinking, at a time of great political and strategic change. In sum, while the overall size of U.S. land forces has been dramatically reduced, the proportional size of the institutional base has been reduced even further. If, in Iraq and Afghanistan, the field forces have been stretched to their limits, the institutional army and Marine Corps at home have been stretched further still. We have been maintaining a punishing tempo of operations at the expense of the institutional base's health, thus running a strategic and a general long-term risk.

Information Gathering and Processing

At the same time, the demands of the military profession have been increased by the many complexities—tactical, operational, strategic, political, and cultural—of the Long War. As one senior Iraq veteran recently wrote, "The U.S. military is at a turning point in its history. We risk losing our status as the greatest military power in the world if we do not rapidly institutionalize fundamental changes in the way we train, fight and lead."[3] That is, the institutional services must generate not only larger land forces but better land forces. The good news is that doing so builds upon an enduring strength—the institutions of the army and Marine Corps already excel at producing leaders; but the challenges, both quantitative and qualitative, are significant. Force planners must adjust their measures to include the institutional basis of the services, that is, the "headquarters" elements that are the key to generating and sustaining forces to be employed in operations. The U.S. Army and Marine Corps must be able to maintain large and powerful forward-deployed and forward-operating forces twenty-four hours a day, seven days a week, 365 days a year, in an increasing number of austere operating environments far from their home stations, without respite for the foreseeable future. This is a tremendous strain on the field force (that portion of the army and Marine Corps organized into brigades and regiments) but also on the institutions of the services, which must generate sufficiently numerous and sufficiently savvy forces to succeed in this more complex combat environment.

Again, these new demands turn accepted "transformational" tenets upside down. For example, one of the great technological promises of recent years was that the battlefield would be increasingly transparent to U.S. forces. In the mid-1990s, Adm. William Owens, then vice chairman of the Joint Chiefs of Staff, described the prospects for "dominant battlespace knowledge," under which "the increased density, acuity and connectivity of sensors and many other informational devices may allow U.S. Armed Forces to see almost everything worth seeing in real or near-real time."[4] This "system-of-systems" approach would "build the realm" of "near-perfect force" allocation:

> We will increasingly assign the right mission to the right force, matching our forces to the most successful course of action at both the tactical and operational levels of warfare. Further, our increasing capacity to use force faster, more accurately, and more precisely over greater distances and interacting with the advances in [intelligence, surveillance, and reconnaissance capabilities] will build a qualitatively better realm of battle assessment. We will know the effects of our actions—and understand what those effects mean—with far more fidelity, far earlier than anything we have experienced to date. This dominant knowledge, in turn, will make any subsequent actions we undertake even more effective, because we will truly be able to operate within the opponent's decision cycle, and the opponent's capacity to operate will have been greatly eroded.[5]

But as suggested in the previous chapter, the experience of land warfare in the post-9/11 period has frustrated nearly every aspect of the transformational approach. The hope to assign perfectly tailored forces to Afghanistan and Iraq has underscored the inherent brittleness and fragility of such a strategy. Believing so deeply in our own ability to see the battlefield, we have failed to see our own misperceptions. Our strategic situational awareness has become, as former defense secretary Donald Rumsfeld might have put it, an "unknown unknown." If there is a single quality that U.S. land forces must recover, it is the ability to operate—and to win—in an uncertain and opaque environment, accepting that perfect clarity is unattainable.

It remains true, however, that the Long War is first and foremost a struggle for and about information, and U.S. land forces must now principally be organized around the need to gather, analyze, share, and act in response to a flood of information. Long War veterans understand that they must be prepared to operate in the realm of information. Indeed, the experiences of Iraq and Afghanistan have led many to the conclusion that traditional, hierarchical military organizations are inimical to the necessary flow of information. Galula's first principle was that a counterinsurgent must first arm himself with a cause; a modern corollary might be that he needs to arm himself with a narrative.[6] The *jihadi* enemy, in particular, is adept and adaptive, weak in firepower but strong in "information power." As one officer observes, "The enemy uses the same information available to us, and his decentralized organizational structure and unconventional operations complicate our ability to predict his actions."[7]

The enemy's unpredictability highlights the need to refocus land force intelligence efforts, structures, formations, and systems. Current intelligence systems have long been structured to identify conventional enemy formations and to predict their actions based on relatively transparent past patterns, known tables of equipment and organization, and published doctrine; they sought predictability and enshrined the concept in the idea of "intelligence preparation of the battlefield." But, as one brigade commander put it, "Exactly none of these conditions existed after Saddam's army was defeated."[8] Such conditions never existed in Afghanistan; even the most elemental facts, such as the Taliban order of battle, were not clear. Nor are conventional systems or methods well designed for such environments: imagery, electronic surveillance, standard combat patrols, and techniques of scouting are of limited value in a counterinsurgency. Continues the brigade commander, "We neither understood nor anticipated the inadequacy of our conventionally designed intelligence collection and analysis system. More importantly, almost no one understood the dominant role that [human intelligence] operations would play in developing actionable intelligence on a burgeoning insurgency."9

Predictably, the process of collecting and, equally, of analyzing human intelligence, or HUMINT, is a manpower-intensive effort. The brigade commander quoted above effectively doubled the size of his tank unit's intelligence organization, and one suspects that there were concomitant increases in his subordinate battalions. "The amount of information that must be collected, analyzed and synthesized to produce actionable intelligence can be overwhelming," he wrote:

> Personnel needed for activities such as document and technical exploitation, interrogations, informant meetings and plans and current operations present additional manpower challenges . . . The number of authorized billets and Military Occupational Specialties is simply inadequate to conduct and sustain HUMINT-centric operations . . . We had more than enough combat power in our organizations to overmatch the enemy in Iraq; what we didn't have was the depth and knowledge in our intelligence sections to find the enemy in the first place.10

Operations such as those in Iraq and Afghanistan demand that leaders stress intelligence as a central mission of combat units: this brigade commander placed a high priority on generating his own intelligence, personally attending and leading a weekly "reconnaissance and surveillance" meeting. The intelligence process itself in irregular warfare comes to resemble a criminal investigation, so that the primary source of good information is local informants. This situation requires intelligence-gathering methods vastly different from traditional ones, and imposes challenges in finding, vetting, training, equipping, and evaluating human sources. In Iraq and Afghanistan, units have routinely benefited from "walk-in" HUMINT sources; the problem is to evaluate and develop a more complete picture and to make the most of more reliable sources. Units often provide Global Positioning System devices, digital cameras, cell phones, and Internet access to sources in order to get better intelligence more easily from them and to reduce their risk of exposure. But in each case, these improved methods demand complex and time-consuming training for sources.

There is also a problem in exploiting the successes of HUMINT-driven intelligence, such as when enemy computer hard drives or cell phones are captured. Tactical units need the ability—mostly, the software and training and the linguists—to exploit the contents of captured computers, which can be extremely helpful in understanding the workings of local insurgent networks or in building targeting information. Similarly, many units, particularly in Iraq, are frustrated by their inability to exploit detainees. Some of this is simply a problem of detainee capacity, but other issues include the preparation of evidence (because detainees must be turned over to local legal authorities, intelligence gathering must be shaped to meet not only the immediate needs of U.S. or coalition military operations, but also judicial standards), the ability to infiltrate detainee facilities, and the ability to correlate lines of interrogation with a unit's intelligence needs.

In sum, the demands for intelligence assets at the tactical level have mushroomed. Even as, during the "surge" in Iraq, increased patrolling and overall troop presence have generated an increased volume of information, the need to understand, analyze, and exploit that information has likewise risen. Thus the army's deputy chief of staff for intelligence envisions that intelligence sections of battalion and brigade staffs will continue to grow.[11] In addition to the growing number of soldiers and marines devoted to human intelligence activities, there is a growing emphasis on language and cultural skills. For example, in 2005 the Marines established a Center for Advanced Cultural Learning to provide predeployment training to units at all levels, to integrate such training into professional military education, and to institutionalize cultural awareness training across the force.[12] Cultural awareness training efforts now begin in basic training.[13]

The need for culturally conscious intelligence underscores the centrality of another aspect of the information battlefield. In the Long War, and following Galula's precept, U.S. land forces do come armed with a cause—the goal of bolstering stable, representative, legitimate, and allied governments in the Muslim world—but they are far from perfectly equipped for their mission within the broader cause. To be sure, many of the tasks of shaping and advancing the liberal and modern alternative to the *takfiri* narrative fall to other agencies of the U.S. government, but the military, and land forces in particular, have their role to play. The art and science of "information operations" is not well developed, and it is misleading in any case to think of them as somehow separate from other kinds of operations; but the result is to impose another requirement on the service institutions. Soldiers and marines must not only be aware of the cultures in which they operate, but must understand something of how to translate our goals into a foreign frame of reference; they must not only destroy enemy military formations, but contribute to the transformation of a hostile political culture.

This mandate adds yet another level of complexity and opacity to the operational environment for U.S. land forces. It further indicates that it is all but impossible to attain the transformational ideal of a transparent battlefield, which means that the ability to perfectly tailor forces—at the strategic as well as the tactical and operational levels— is equally uncertain. Precisely designed forces are likely to be precisely wrong. Land forces must again be prepared to pay the bills of uncertainty. Thus the first measure of this return to more robust force planning needs to be a renewed emphasis on generating a sufficient reserve. This is primarily a numbers problem, one to be discussed more thoroughly in the following chapter. But there are other, qualitative requirements that can contribute toward making U.S. land forces more flexible, more robust, and better prepared to operate in uncertain environments.

Firepower

In thinking about these requirements, it would be wise to revisit one of the central tenets of the transformation movement: that a transparent battlefield allowed for not only an overall economy of force but an economy of firepower, and, more specifically, that precise airpower could often substitute for artillery fire support. While this insight was a powerful one, it was, like much of the transformation project, premised on an understanding of warfare that is only occasionally reflected in recent experience. The difference is evident in the case studies explored in chapter 3: the tactics that produced the stunning results during the invasion of Iraq are not directly applicable to the battle for Fallujah, to take but one example, and are even less so during day-to-day counterinsurgency operations.

A report done by RAND's Project Air Force in 2005 captured the trend of declining fire support in land operations.[14] Looking at a range of operations going back to the invasion of Grenada in 1983 and assessing the structures of units emerging from the initial efforts at defense transformation, the study concluded that "the weight of fire support allocated to brigades participating in major combat operations appears to be declining over time." Further, the historical trend was bound to continue as new unit designs transformed what had been a choice into an imperative. The study's estimates were that the army's new "modularized" brigade combat teams "will have only 25 percent of the fire support provided to heavy brigades in 1991." RAND rightly concluded that "this 75 percent reduction will be Army Transformation's most important consequence, particularly for the Air Force."[15] The study went on to make a number of recommendations to improve cooperation between the air force and army, suggesting that the air force increasingly develop a "counterland" orientation.

But it is not at all clear that closer cooperation or improved precision can fully make up for the loss of organic ground-force fire support. As air force Lt. Gen. Gary North, the regional air commander in Iraq, told National Public Radio, the current tactical situation does not lend itself to the heavy employment of airpower. "In Iraq you've got a very mobile enemy that moves around in twos, threes, fours, fives, 10s," presenting a very small target.[16] Perhaps more importantly, the target is fleeting: the problem is not so much the volume of fire as the availability of fire. As with other elements of land power, the most salient requirement for fire support in the emerging battlefield environment is persistence and ready availability, the qualities that are most problematic for airborne fire support, whether from fixed- or rotary-wing aircraft. An after-action report from the second battle of Fallujah—where, because of the time available for detailed planning, air support was more plentiful than it is under "normal' conditions—concluded that howitzers

> were able to provide timely and accurate fires throughout the fight, delivering 925 rounds in mostly danger close fires (that's less than 600 meters from friendly soldiers and often within 100-200 meters from friendly forces). A big lesson learned is that, even when responsive, close air support is not a substitute for artillery and mortars. It can be very effective, but it is not as responsive as our artillery and mortars.[17]

Those who commanded the Fallujah operation came to similar conclusions. About six thousand artillery rounds were fired, all of them in response to insurgent actions; there were no general suppressive fires, nor were harassing and interdiction fires shot without specific targeting information. Crucially, every round was controlled by a forward observer, suggesting that the limited number of air force ground controllers (and their relatively short tours of duty, averaging just four months in Iraq) is another problem in providing fixed-wing fire support in long-running operations.[18] Finally, the dispersed nature of U.S. land combat units in counterinsurgency operations multiplies the need for responsive fire support. The volume of fires was not that high at any one time during the second battle of Fallujah—the two brigade-sized marine-army ground maneuver task forces in Fallujah each had but a single battery assigned in direct support—but the need for constant support over twelve days of intense fighting accounted for the large total of fires.

The need for responsive indirect fires is especially important in light of the trends in land force modernization. As the army's force structure becomes proportionally lighter, with a growing number and ratio of light infantry and medium-weight Stryker brigades and with the plans for the Future Combat System ground vehicles, its direct-fire capacities are diminishing. A heavy unit equipped with Abrams tanks and Bradley fighting vehicles continues to enjoy a firepower advantage over most likely adversaries, but smaller and lighter vehicles naturally entail some greater risk. As the army and Marine Corps reshape themselves for the future, it would be wise to reevaluate these risks.

Leader Training

Above all, what transforms numbers and capabilities into results is the quality of leadership. The performance of the U.S. Army and Marine Corps in the occupation of Iraq and Afghanistan has been mixed, although the quality of tactical leadership has surpassed the quality of generalship. The Long War and, arguably, the emerging nature of land warfare in the twenty-first century, place new stresses on junior leaders: both the quantity and importance of their decisions have been increased. Consider first the pace of decision making, a direct result of dispersion and pace of operations. "The environment we faced required junior leaders to make hundreds of independent decisions every day," recalls a cavalry squadron commander. "The sheer volume of information generated daily was staggering. Moreover, the operations tempo was very high, requiring the execution of dozens of missions simultaneously across the spectrum of operations."[19]

Then there is the quality of the decisions forced upon junior leaders. Marine Corps leaders have long spoken about "the strategic corporal," suggesting that even the lowest-ranking NCO, in charge of a handful of fellow marines, might find himself in a situation where his actions, or lack of action, have genuinely strategic effects, especially if magnified through the lens of an omnipresent and omnivorous international press.[20] We have asked our military leaders not only to fight an unexpected kind of war, but also to make decisions far outside the scope of their training: to act as mayors of cities, to supervise public works projects, to reform local politics, even to conduct diplomacy. Officers and NCOs with but a few years' military experience are thrust into an environment for which their training is, at best, merely illustrative. As the cavalryman, a commander in the successful missions in the Iraqi border town of Tal Afar, wrote, these soldiers and marines must have "critical thinking skills so that leaders at all levels have not only the knowledge and training, but also the judgment, to make the right decisions in a combat environment."[21]

Thus military leadership training must move beyond simple instruction to become more a matter of education. "Good basic skills alone will not guarantee success," writes one officer.

> We must take our soldiers out of their comfort zones from the first day of training by replicating the fog of war even during rote tasks . . . Many units and schoolhouses continue to train using rote methods only, fostering a checklist mentality among those who receive the training . . . We must place them in training situations that consistently generate overwhelming amounts of information.[22]

Even if other agencies of the U.S. government begin to assume some of these burdens outside the scope of traditional combat, many of them will still devolve to men and women in uniform. This reality reinforces the notion that land force leaders must be more fully educated as well as trained. Obviously, they must learn new languages and understand diverse cultures, but they must also acquire a more sophisticated strategic and political understanding, since the acts and decisions of junior officers and noncommissioned officers can quite easily have strategic effects. And, as suggested above, their mission is not simply to operate in a foreign culture, but to have an effect upon it. There is no way to provide land force leaders with such skills except by improving their training and professional education. Improved training and real education mean time, people, and money. Even the most successful training programs of the past decades are now insufficient. "Training to the technical standard is essential," concludes the cavalry officer, "but I believe it is equally important to view the training of these tasks as a vehicle to cultivate judgment in your subordinates." Even home-station, small-unit training, for example, "should include the complications of dealing with civilians in a combat environment."[23]

Leaders with experience in Iraq and Afghanistan describe a difference, as one put it, between the past system of "command-centric" decision making and the emerging habit, offered as the model for future training, of "soldier-centric" decision making.[24] Historically, the goal of leadership training has been to give structure to decision making in the chaos of combat, the fog of war. Leaders formulated plans, sought predictability, focused their efforts on producing clear orders that set forth defined tasks. Professional education provided a punctuation to field assignments. It was designed to improve decision making in combat and was either service-specific or, in recent years, joint-service. Matters of strategy were withheld until the war college years, reserved for senior colonels and generals. Today and going forward, good leaders need to be self-aware and adaptive; they must impart to their subordinates their intent and an understanding of the mission, rather than a highly specific plan and orders. They must be comfortable with uncertainty and ambiguity and create the conditions where subordinates are free to exercise initiative. Leaders at lower levels must be prepared to deal with matters of strategy and with the second- or third-order effects of decisions, and they must understand the strengths and weaknesses of the other agencies of the U.S. government and even international nongovernmental organizations. Education must be a constant process, with formal professional military education designed to broaden horizons rather than to teach "schoolhouse" solutions.

Partnership

A further emerging requirement for U.S. land forces for the future and particularly in the conduct of the Long War is the ability to partner with allies; as the report of the 2006 *Quadrennial Defense Review* put it, "Helping others to help themselves is critical to winning the Long War."[25] This forward-looking trend, often described as "building partnership capacity," is an immense challenge, and immeasurably different from the kind of mature alliance relations that marked NATO or other partnerships of the late Cold War. Building partnership capacity can mean, as in Iraq and Afghanistan, creating national military establishments entirely from scratch— and indeed, with even a prior step of destroying the previously existing establishment. Further, it can mean building other elements of the partner state, particularly other elements of the security structure, including national and local police and even a judiciary. Finally, while other agencies of the U.S. government can and should play important roles in these efforts, it is inevitable and in a number of ways preferable that the military, and U.S. land forces especially, play the largest part. As retired Lt. Gen. David Barno, the commander of U.S. forces in Afghanistan from 2003 to 2005, recently wrote: "My guidance to our staff was that as the most powerful organization in the country, we would take a direct interest in *everything* [italics in original]—not just the traditional warfighting piece. As I told an exasperated and overworked staff officer in October 2003: 'We own it all!'"[26]

The partnership-building mission is central to victory in a long and irregular war, where the strategic center of gravity is less the enemy than it is ourselves and our friends; successful counterinsurgencies must first protect the people, and destruction of enemy formations is a means, and a secondary one at that, not the end in itself. The emerging recognition of the full extent of the partnership-building task is a reflection of bitter experience and repeated failure, especially in Iraq. When, in May 2003, the Coalition Provisional Authority in Iraq dissolved the various "entities" of the former Saddam government, it committed the United States, quite unintentionally, to what is almost surely the largest and most complex task of advising foreign military forces in our history. "With the formal disestablishment of Saddam Hussein's Iraqi military forces, U.S. and coalition forces were suddenly faced with the daunting task of rebuilding an Iraqi army from scratch," says Army Col. Frederick Kienle, one of the officers assigned in 2004 and 2005 to the effort. The Iraqi Army was one of the few respected or even partially functional institutions of the Iraqi state, but it was also a feared tool of Saddam's repressive rule. "Its few vestiges could neither be reconstituted nor rebuilt around what little leadership and structure could be salvaged."[27]

The rebuilding of the Iraqi Army was misconceived from the start. The guidance of the Bush administration—"When the Iraqis stand up, we can stand down"—either misunderstood or misrepresented the fact that creating a decent post-Saddam Iraq demanded a long-term U.S.-Iraqi alliance, and so the army-building exercise began from faulty premises. The original conception, which was formulated even before the U.S. invasion, was that the Iraqi Army would be a small, weak territorial force that would not threaten Iraq's neighbors. In other words, it would be designed in an effort to avoid repeating the experience of the Iran-Iraq War or the invasion of Kuwait in 1990. The job of training the new force—in a sterile, classroom environment, and even across the border in Jordan—was given to contractors. But as the Sunni insurgency grew through 2003, the inadequacies of this approach became increasingly apparent, and during the first fight for the town of Fallujah, in 2004, it collapsed entirely when one Iraqi unit refused to fight and others performed miserably.[28]

A renewed training program began in 2004, with an additional component of embedded U.S. advisors in Iraqi units. While this program slowly began to produce better Iraqi units, the new Multinational Security Transition Command–Iraq, or MNSTC-I (pronounced, inevitably, "Min-sticky"), was manned in an ad hoc fashion, with a large proportion of reserve and Army National Guard officers. Moreover, the mission of "foreign internal defense" (the doctrinal term for building partnership capacity) had long been the purview of army Special Forces, and Green Beret units were heavily committed to other missions. With time, new "military transition teams" have proved more useful in building stand-alone Iraqi and U.S.-Iraqi effectiveness. The army has redesigned one of its tank battalions, First Battalion, Thirty-fourth Armor, at Fort Riley, Kansas, to prepare these teams for deployment, and the Marine Corps is establishing a similar unit.[29] These teams are also getting more extensive field training at the army's Joint Readiness Training Center. But the success of the mission remains hampered by the limits of the "transition" mission. In fact, the most successful partnerships, measured both by immediate tactical performance of Iraqi units and by progress toward the self-sustaining capacity needed in future, have resulted in the change of tactics associated with the recent "surge" in Iraq. The more active posture of U.S. units, remaining in their areas of operation for extended periods rather than "commuting" from large forward-operating bases, is as important in improving Iraqi performance as the embedding of training or advisor teams. Successful partnerships, it should come as no surprise, are the fruits of close and long association, not of a rush to "transition."[30]

Partnership building in Afghanistan, while hardly perfect, has always seemed to be a step or two ahead of similar efforts in Iraq. To begin with, the Bush administration more readily accepted from the start the necessity for a long-term commitment in Afghanistan, though there was and remains an unresolved tension between the counterterrorism mission to chase down al Qaeda's senior leadership, and the counterinsurgency mission to protect the Afghan people and build the Afghan state. An Office of Military Cooperation was established in mid-2002, shortly after the invasion, to begin building an Afghan National Army (ANA). After something of a slow start, this effort was accelerated and expanded during Barno's tenure. As he writes:

> Despite intense tribal rivalries, the ANA and [Ministry of Defense] were re-created with an ethnically balanced, merit-based leader selection process that, by late 2003, had established both as models among the most-reformed bodies of the Afghan government . . . The strikingly positive reaction these [ANA] units evoked when they entered villages alongside their embedded

U.S. trainers stood in stark contrast to the reactions
elicited by the repressive tribal militias . . . Their professionalism, discipline, and combat effectiveness stood
out; they became sources of national pride.[31]

From 2003 to 2005, no ANA unit was defeated or broken in a
combat engagement—testimony to the success of the training
program conducted by then major general Karl Eikenberry (who
succeeded Barno in overall command) and to the fact that whenever major ANA units were sent into harm's way, they were accompanied by embedded advisors and were backed up by U.S.
reinforcements and on-call firepower. Unfortunately, the "transition" virus also infected the Afghanistan mission, driven by the
Bush administration's desire to draw down forces, to shift the burden of the counterinsurgency to NATO allies, and generally to
declare a victory as things turned to the worse in Iraq. But beginning in late 2006 and continuing into 2007 there has been both a
slight surge of U.S. forces and a recommitment to the mission, in
part driven by the revival of Taliban strength; and new efforts are in
train to complete the building of the ANA and to extend reforms to
the Afghan National Police. Significantly, the U.S. military has
found it necessary to take a larger role in police reform. But as
Barno concludes, "With the police widely acknowledged to be the
'first line of defense' in a [counterinsurgency] campaign, it remains
unfortunate that the fusion of police and ANA training oversight
came so late."[32]

    The challenges of partnership in Iraq and Afghanistan represent
one extreme of this new requirement for building partnerships. An
equally important element is suggested in the Philippines case
study of the previous chapter: there is a need to work preventively
with indigenous forces before a crisis or war requiring a major
commitment of American forces. While in most cases the scale of
these efforts will not reach that of Iraq or Afghanistan, there will be
many places where lower-level but long-term partnerships must be
built or buttressed. Given the centrality of these missions, it is
crucial to think through what qualities U.S. land forces will require

to conduct them, and then to translate those general qualities into force-planning measures. Many of these specifics will be discussed in greater detail in the chapter below, but the experience of recent years makes clear a number of general requirements.

While the ability to create a partner military from scratch is the extreme case, it forms a necessary complement to the requirement for thoroughgoing regime change. That is, in those rare but necessary instances when U.S. forces are called upon to invade a country and change its government, we must expect to undertake a large and long process of military and broader security-sector reform and reconstruction. Just as American land forces must retain a certain ability to knock down rogue regimes, they must retain an equally sure capacity to erect a secure successor regime—a long-term strategic partner. Just as there must be ready, deployable combat forces available for these missions, there must be similarly expeditionary advisory units. If there is one lesson from Iraq and Afghanistan, it is that, even in a Long War, time is precious.

Second, we must enhance our partnerships with a host of strategically important governments and militaries across the Muslim world, throughout the battlespace of the Long War. Doing so will involve complex and diverse tasks, varying widely according to local conditions. A rapidly deployable advisory corps is not well suited to this sort of effort, where much depends on years of engagement and working with local governments, which are often taking the first tentative steps toward democracy and social reforms. As a captain, a two-tour Iraq veteran, recently wrote in the *Small Wars Journal*, "[N]o matter how good the training, nothing can match the value of actual advisor experience when it comes to training and advising another human being in a cross-cultural environment."[33] In such cases, a better model is to be found in the military advisory and assistance group, integrated as part of the all-agency "country team" led by the State Department chief of mission, and providing the continuity and length of experience needed to create durable partnerships.[34] While these may be relatively small contingents (though very senior by rank), they are flexible and can be expanded dramatically in a crisis. In 1949, almost a year before the Korean War started, and despite the fact that Korea would be "excluded" from the American security perimeter, the Truman administration established a military advisory group to the Republic of Korea, known as the KMAG, whose job it was to improve the quality of the Korean Army. The next year, a similar military advisory and assistance group was established in what was then known as Indochina.[35]

Expansiblity

There is one final quality, beyond sustainability and robustness, lethality and leadership, beyond the need to build new alliance partnerships, that must be reemphasized in the rebuilding and reshaping of American land forces, and that indeed cuts across all these traits: our land forces need to be genuinely expansible. We must accept that the state of the world is now one of great uncertainty, and that this has implications from the level of tactics to that of grand strategy. Already, the size and duration of the land force commitments since 9/11 have come as a nasty surprise; we must accept that the pace of events could well accelerate and that our adversaries will present unpredictable challenges. We must design our forces to rise to those new challenges in a timely fashion.

Alas, the quality of expansiblity has been almost lost by the decision-by-default to fight the Long War with too small a force. For five years, activated reservists and National Guardsmen have been providing 15 to 20 percent of present U.S. Army active strength; they are no longer a strategic reserve, a hedge against unforeseen contingencies, but an operational reserve, as consumed in their own fashion by the rotational demands of Iraq, Afghanistan, and other Long War efforts as the regular force is. And their equipment stocks have been looted in an even greater measure. In sum, the Bush administration's failure to expand, refit, and restructure U.S. land forces in a timely fashion after the 9/11 attacks has left the army and Marine Corps dangerously brittle. The spirit of soldiers and marines is undiminished, and their performance in battle has been superb. The force is not broken, but its institutional basis is cracking.

The fact is that the army and Marine Corps are so tautly stretched that they are poorly prepared to accomplish the growth they so desperately need. In a challenging recruiting environment, the services are hard pressed to increase the number of recruiters. Even if a flood of new soldiers and marines was attracted to serve, it would be very, very difficult to train, organize, and equip new units at a more rapid pace. Barracks, bases, ranges, and equipment stocks have been "streamlined" relentlessly over the past fifteen years; the wisdom of the base realignment and closure processes now appears very uncertain. So as U.S. land forces struggle to grow themselves—itself a process that is likely to take a decade—they must reacquire the suppleness and flexibility that would allow them to grow again should the need arise.

The professional, all-volunteer force of the late Cold War years was well suited to its tasks: punching well above its numbers in a highly lethal form of conventional combat against a Warsaw Pact onslaught, until waves of reinforcements could arrive. Strategic success was measured in weeks—even days, if one remembers the ten-divisions-in-ten-days benchmark for deployment to Europe—and failure by the possibility of nuclear Armageddon. The new tasks could hardly be more different: combat is highly irregular, and may matter less than stability operations; regulars man the perimeter with little prospect of a national mobilization; success is a matter of decades, if not generations; and still failure might be marked by the employment of weapons of mass destruction, not in waves of rockets but in a repeated drip of suicide attacks. American land forces are less a wall of steel than a sponge, absorbing *takfiri* toxins and leaching them from the lands they incubate.

Indeed, the qualities enumerated above all reflect the realities of the Long War. As the case studies of the previous chapter suggest, we have accumulated sufficient experience in this conflict, much of it bitter, to abstract out the characteristics of a force that might be successful on these new battlefields. We began the fight with the force on hand, but it is past time to begin the process of creating the force we need.

5

Costs: Time, People, Money

The United States requires a million-man active-duty land force, a properly balanced mix of marines and soldiers, configured in such a way as to win the Long War. This is the primary purpose of the land force. Of course, it is not the only purpose: the services must retain a sufficiently varied set of capabilities to be able to respond successfully to a broad spectrum of conflicts. Nor is success in battle the only test of land forces or of the army and Marine Corps as institutions. To meet the nation's needs, we must look beyond the number of active brigades and regiments. We must restore the reserve components as a genuine strategic reserve, not simply another pool of resources to meet immediate rotational requirements. Most profoundly, we must invest much more in the process of training, equipping, and educating our land forces.

A Ten-Year Commitment

Rebuilding U.S. land forces to the required level is the work of a decade. The current total active army and marine strength is slightly less than 700,000 men and women. The plan advanced by Defense Secretary Robert Gates calls for a very modest increase, to about 750,000, over six years, a rate of increase of about 8,300 per year. Bringing the army back to slightly above its 1988 level, to approximately 800,000, and enlarging the Marine Corps to 210,000, slightly above the level anticipated under the Bush administration's plan, would require 30,000-per-year increases over ten years. While this might seem an excessively modest pace—and every effort should be made to accelerate it—it nearly quadruples the rate of

110

increase enshrined by the Bush administration. And though the level of troop retention through the difficult years of Iraq and Afghanistan has remained remarkably high, given the stress on soldiers, marines, and their families, attracting new recruits has been a challenge. A commitment to expansion, with the concomitant understanding that a larger force is likely to be a less frequently deployed force, might well contribute to eased recruiting, as, even more, might a call for service on the part of American political leaders of both parties. But a realistic assessment makes clear the need for a long-term effort.

The other elements of a proper expansion are likewise time consuming. Consider the increased training requirements of a larger force. The post–Cold War drawdown and the mania for "overhead" efficiencies has resulted in a diminished training base; even with the Iraq-era supplemental appropriations, the first priority for dollars, people, time, and facilities has been the field forces. Army leaders have often indicated that the lessened "throughput" capacity of their training pipeline is an equally constraining factor for expansion. Indeed, one of the first priorities of expansion should be the restoration of the training base so as to make further and faster growth possible.

Equipping a larger force will be a similar challenge. Even if one believes that the most pressing need is for dismounted infantry, the land force industrial base has suffered from fifteen years of disinvestment. The United States has not manufactured a new Abrams tank or Bradley fighting vehicle or marine Light Armored Vehicle for many years, and the unexpected intensity of fighting in Iraq has resulted in a tremendous reduction in the prior inventories of equipment of all kinds. Again, recent supplemental appropriations have helped alleviate some of the repair backlogs; yet many of the affected vehicles were already old. Remanufacturing can extend the life of these systems, which are still dominant despite being designed in the 1970s and introduced into service fleets during the Reagan defense buildup; an M1A2 model tank is vastly more capable than the basic Abrams introduced a generation ago. Yet simply repairing the old gear, good as it is, is no substitute for real equipment modernization. As will be argued below, the army and marine modernization programs first conceived in the 1990s were similarly starved of funds and have proceeded at a too-slow pace; major investments will be needed to accelerate these new systems, as very few of them have reached a state that would allow for high-rate, serial production. To extend the life of old systems is to lock land force units into the doctrine and tactics of the past. Modernization efforts must be pursued with a sense of wartime urgency.

Most time consuming of all will be the education of the force to meet the demands of the emerging regime of twenty-first-century warfare—highly irregular and asymmetric yet highly technological and intensely violent at the point of contact with the enemy. While much commentary has focused on improving language skills, cultural awareness, and other capabilities that would improve the tactical situational awareness of soldiers and marines in combat, an equally crucial component in future success is injecting a sense of what might be called "strategic situational awareness" at even the lowest levels of operation. The last chapter alluded to what marine leaders call "the strategic corporal." The phrase suggests both that the actions of even the most junior enlisted leaders could have large-scale, strategic effects, and that small-unit leaders may be seriously underprepared for this responsibility. Marine grade corporals (or their army E-4 equivalents) may have just a few years in service; they are often younger than twenty years old, and their mastery of weapons and tactics is extremely basic. They are well disciplined, bright, and inventive; yet there is little in their military training, beyond field exercises and previous experience in battle, to instill the level of knowledge or judgment they may need.

What is true at the lowest level is distressingly true at more senior echelons of command. While the officer corps has, in response to the crises and challenges of Iraq and Afghanistan, done a remarkable job of self-education, much of this was originally, and remains, inspired improvisation. An important exception was the rewriting of the joint army-marine counterinsurgency manual, but the shifting sands of the Long War will require a constant reevaluation of doctrine and strategy; the imperative to think about war is as strong as the imperative to fight it. A commitment to military education, along with the acknowledgement that time was a key to deep reflection and true understanding, underpinned the overall land-force reforms of the late Cold War. This was a necessary predicate for the development of doctrine, equipment designs, and operational concepts that made the army and marines such masters of late-twentieth-century maneuver warfare. The unique forms of warfare that mark the twenty-first century deserve no less study and reflection.

Sizing the Force

Throughout the Cold War, the essential defense planning question was, "How much is enough?" In the 1990s, in the immediate aftermath of September 11, and during the invasions and occupations of Afghanistan and Iraq, determining the proper size of the U.S. military was plagued by uncertainty over what the mission was; the first-order question was, "What do you want us to do?" The previous chapters of this report have attempted to narrow this elemental uncertainty to the point where normal "force-sizing constructs" can be employed, at least in regard to U.S. land forces. And, as argued repeatedly above, the central concern, driven by the centrality of the Long War mission, is to determine the appropriate size of the active-duty component of the U.S. Army. All the other force-sizing questions—regarding the Marine Corps and the reserve components of the army—depend upon the size, structure, posture, equipping, and, not least, cost of the active army.

The first principle of land-force sizing should now be the need to conduct a sustained, large-scale stabilization campaign, either in lieu of or after a regime-changing operation, without so stressing the force that it is incapable of responding to another crisis or crises, including the initial phases of a major combat campaign. Ideally, the active-duty army alone would have the ability to meet such a commitment: the mission places a premium on long-term professionals, that is, active-duty regulars; and it demands the kind of sustained presence that is the unique and core competence of the U.S. Army. This level of land force capability should be maintained no matter what happens in Iraq; this is not a surrogate argument about Iraq policy. Already there is an anti-Iraq strain in the Pentagon—"Let's never do *that* again!" But that is exactly the wrong conclusion. The decision to invade and reconstruct Iraq was taken after all other options were tried and found wanting; a dozen years passed from the beginning of Operation Desert Storm to the fall of the Saddam statue in al Firdos Square. We may be equally anxious to avoid an occupation-type mission in another large country, but it would be negligence to plan as though it could never happen. There is a significant number of large-yet-volatile states whose stability might be a crucial U.S. interest—Iran, Pakistan, Nigeria, or even a post-Communist North Korea, to name but a few.

To repeat: this is not a prediction of conflicts to come, but a recognition that the potential for stabilization and reconstruction missions remains high. We may not want these missions, but they might be thrust upon us; and they certainly might appear to a future president as the least-bad outcome. And, thanks to hard-won experience, we can quantify this requirement. Roughly speaking, the Iraq reconstruction mission has required a steady force of about eighteen brigade combat teams (BCTs)—the somewhat unfortunate yardstick by which the army now measures things—along with a healthy measure of "enabling" units, and thus force size hovers between 130,000 and 160,000. Future individual cases might require more or fewer troops, but an "Iraq-sized" block remains a basis for building U.S. land forces.

A second set of blocks should be sized on the ability to conduct several less-demanding reconstruction missions simultaneously. Call these "Afghanistan-sized" blocks. While Operation Enduring Freedom receives far less press attention than Operation Iraqi Freedom, our experiences in both have in many ways been broadly similar. The former has been a long mission—indeed, longer than Iraq—and it demands a significant force. In fact, despite the commitment of a NATO force of about sixteen thousand and the more successful and more rapid reconstruction of the Afghan National Army, U.S. troop levels in Afghanistan have remained relatively steady, hovering between twenty-five and thirty thousand in recent years, including three to four brigade combat teams. Again, whatever

the direction of U.S. policy over the coming years, we would be fool-
ish to plan for a withdrawal from Afghanistan, particularly with the
situation in Pakistan appearing increasingly unstable and the presence of the Taliban and al Qaeda senior leaders just across the border.
A more sensible construct would be to plan for a reinforcement (on
the not-unreasonable assumption that our NATO allies might reduce
their commitment, a commitment they have never fully met as it is)
or the prospect of a second such mission. And so there should be two
"Afghanistan-sized" blocks in the foundation of U.S. land forces, one
in the active army and a second in the Marine Corps.[1]

A third set of force-building blocks should be loosely modeled on
the Combined Joint Task Force–Horn of Africa, or CJTF-HOA. The
task force is, as former head of U.S. Central Command Gen. John
Abizaid put it, a "blueprint" for one kind of partnership-building
mission. "Dollar for dollar and person for person, our return on our
investment out here is better than it is anywhere in the CENTCOM
[area of responsibility]," he has argued.[2] The task force was established in October 2002 with about four hundred members from all
four services; its commander was then–major general John Sattler,
the marine who went on to command forces in the second campaign
in Fallujah. Key to the initial effort were the command and control
capabilities abroad the USS *Mount Whitney*, but by May 2003 an
onshore headquarters had been established at Camp Lemonier, a
renovated former French Foreign Legion post in Djibouti. On average eight to nine hundred U.S. military personnel are stationed
there, along with a roughly equal number of people from other
agencies and contract support groups. And the onshore contingent
often works closely with naval forces patrolling the region; all told,
the task force effort remains fewer than two thousand U.S. forces.
The task force's operational area includes many fragile states (not
only Djibouti, but Ethiopia, Eritrea, Kenya, Tanzania, Sudan,
Uganda, Yemen, and the chaos that is Somalia). As retired admiral
Gregory Johnson has written:

> The mission's focus has evolved over the past four years
> from a "kinetic" orientation to a long-term effort aimed at

[helping] regional governments develop effective gover-
nance and anti-terrorism capacity. It does this through a
series of small scale, capacity-building activities in close
coordination with the respective Chiefs of Mission and
their staffs as well as host nation officials. These include
medical, dental, and veterinarian assist teams; engineering and humanitarian support missions aimed at helping
local officials do beneficial tasks such as drilling wells
and building schools and medical clinics; along with the
traditional military to military training missions.[3]

Again, U.S. land forces should be capable of sustaining up to four
such "partnership" task forces, two provided by the army and two
by the Marine Corps.

A fourth and highly critical set of capabilities—and one where the
full set of joint-service capacities need to be considered and new
ones created—reflects the need to respond to the dangers of nuclear
proliferation. While most commentary on these kinds of missions
focuses on the need to develop or employ strike capabilities to target weapons facilities, there is certain to be a critical role for land
forces in any serious attempt to militarily preempt or prevent the use
of the most dangerous weapons of mass destruction or to render
them safe in a crisis. No mission is more operationally or tactically
demanding. The target sites are hidden, dispersed, and as distant
from potential attack as the proliferating state can make them. Nor
can military planners necessarily count on a long predeployment
period, and it would be a mistake for the United States to allow the
unique and expensive assets needed for such missions to be routinely employed elsewhere. While it may take many years to develop
the capabilities or field the units demanded by this requirement, it
must be taken as a separate force-planning requirement for all the
services, including U.S. land forces.

A fifth set of force-sizing requirements comes from reevaluating
the demands of large-scale, high-technology conventional land warfare (what in past defense reviews was called "major theater war").
One of the essential, if unspoken, tenets of U.S. military preeminence

has been the assumption that, in extremis, a rapid and decisive campaign of regime change could be mounted against all but the largest nations. Yet, in emerging conditions, the invasion of Iraq is a model only of a successful outcome, not a reliable guide to force requirements or sizing. Any such operation in the Middle East, for example, would have to be launched, presumably, from a posture of continued engagement in Iraq and Afghanistan. While in some respects that could conceivably be beneficial—easing the need for strategic deployment, for example—it would also impose limitations. Thus the equal task of the nondeployed force, and a critical reason to raise the readiness of nondeployed units, is to be rapidly prepared to fight an unanticipated large war. This is a simple truth, but it entirely undermines the current army and marine model of rotational force generation and "tiered" readiness. Under the current model, units that return from Iraq, Afghanistan, or any other deployment plunge to the lowest levels of readiness. They turn their equipment in for repair or leave it for replacement units. Their manpower levels dip, often below 60 percent strength; the leadership cadre often is hit harder still. And training, sometimes even at the lowest levels, grinds to a halt. This, in large measure, is why service leaders insist on longer "dwell times" between deployment—not only because soldiers and marines need to reconnect with their families, but because leaders lack the ability to rebuild the units in which they serve in a timely fashion.

This model of force generation has been used by the marines for decades, but was introduced to the army in recent years, along with the concept of "modularized" brigade combat teams. This was yet another "management efficiency" premised on the idea that events would be predictable and that unit rotations could be managed in a brigade-at-a-time fashion. But the result has been inflexibility and incapacity in a time of unpredictability: the so-called "troop surge" in Iraq was a very modest acceleration of the regular rotation plan and generated only five brigades over the period of seven months. Not much of a surge, yet still one that further stressed an overextended force. In many ways, the primary purpose of any program of land force expansion is to repair the unintended consequences of the current force generation models.

By contrast, during the late Cold War, the army expected all its combat units to maintain a high degree of readiness. All eighteen divisions in the active force were maintained at a high degree of preparedness that would have allowed them to deploy and fight at full capacity very quickly. Moreover, the National Guard was regarded as a genuine strategic reserve, a set of combat capabilities to be "surged" if the active forces and their reserve enablers needed reinforcement. Current and likely future demands for land forces require a return to this more supple and less brittle model of force generation, not only for the army but for the Marine Corps, whose Third Division (part of a force structure enshrined in law) has routinely been undermanned and underresourced.

A final, but harder to quantify, force-sizing metric revolves around the panoply of emerging "engagement" missions beyond the traditional theater security cooperation plans of the combatant commands. This is an essential element in the "indirect" or "partnership" approach to the Long War, but longer lasting and smaller in scope than the CJTF-HOA–style partnership mission described above. Little effort has yet been made to translate this requirement into numbers of soldiers and marines (or airmen and sailors, or foreign service officers and members of other U.S. government agencies, for that matter), but it must be a factor taken into account. And while the totals may be small in the overall scheme of things, they will place a significant extra burden on officers and senior NCOs, whose jobs demand experience and education and require the establishment of a training and education infrastructure to support a range of military advisory and assistance groups. Beyond the way these missions are factored into force-sizing calculations, they will demand changes to defense officer and manpower laws and regulations.

And so there are six imperatives that should determine the future size of U.S. land forces: first, maintain the ability to conduct one large-scale counterinsurgency and reconstruction campaign; second, be able simultaneously to conduct two lesser but significant such campaigns, one by the army and one by the marines; third, be able to maintain four large-scale "partnership" missions, again to be divided between the army and marines; fourth, develop new capacities as part of a separate joint-service effort to respond to the problems of nuclear proliferation; fifth, maintain levels of active-duty readiness and reduce reserve component employment to be able to conduct a major theater war without shortchanging any ongoing commitment; and last, broaden the range and scope of longer-term partnership-building missions.

Meeting these commitments would require a total active-duty land force of about 1,000,000, with a reserve component of approximately 850,000 to 900,000, an army reserve of about 350,000, an army National Guard of 450,000 to 500,000, and a marine reserve of up to 50,000. That rough estimate is derived as follows:

- The "major stabilization campaign" building block requires a force of anywhere from fifteen to eighteen brigade combat teams. But it also requires four to five division headquarters, and at least one but far more likely two corps-level headquarters; and in fact, Multinational Force–Iraq really includes two corps-level commands, one army and one marine. Using the traditional three-to-one ratio of total-to-deployed force, sustaining such a campaign requires a base of forty-five to fifty-four maneuver brigades, a dozen or more division headquarters, and up to six corps headquarters. As argued above, the active-duty army needs to be sizable enough to undertake this mission without overly taxing the army reserve or the Army National Guard.

- The second force-building block, the lesser, "Afghanistan-sized" block designed for a stabilization campaign, demands at least three BCT-sized elements, along with a corps-level and a division-level headquarters. As indicated above, a robust land force ought to be able to sustain two of these efforts, one army and one marine, simultaneously with a larger stabilization campaign. Resorting again to the traditional ratios, this block would require an additional nine army BCTs as well as nine marine infantry regiments. This boosts the active-army total to roughly fifty-four to sixty-three BCTs. But it also accounts for the *entire* structure and strength of the Marine Corps *after* the current expansion plans are realized—and thus it underscores the marines' manpower shortages. While Marine Corps force structures are more complex due to their sea basing and reliance on fixed-wing aircraft for fire support, the infantry component is the essential element. It places proportionate demands on higher headquarters elements as well.

• The third force-building block, the "partnership" block modeled on CJTF-HOA, is much harder to quantify, except at the headquarters level. These missions may require relatively few maneuver or combat units, yet they carry great strategic importance and demand a significant staff in support. Thus CJTF-HOA is a two-star, or division-level, command. Both the army and Marine Corps should consider creating additional "cadre" division-level headquarters in the active force as well as in the Army National Guard and marine reserve, not only to lead such ad hoc but long-serving joint task forces but to ensure better integration of all components. Additional, relatively ready headquarters elements would also permit faster and better expansion of the land forces in times of crisis.

• Similarly, the requirements for an on-call "antiproliferation" or "counterproliferation" force are hard to quantify, in large measure because we lack a full understanding of the capabilities needed. Ideally, the command of such a mission would go to U.S. Special Operations Command, but even allowing for the improvements made under the requirement to conduct the global war on terror, it is unclear that USSOCOM has the capacity to plan, command, and conduct such a campaign, one that potentially could include a large number of conventional airborne, air assault, or marine units.

- The fifth building block is based on a return to a force-generation model that stresses readiness across the full active army and Marine Corps. But it also demands that the Army National Guard return to its traditional role of strategic reserve—that is, that it be called out in times of crisis with an expectation of a return to garrison when the crisis is resolved. For example, the guard provided about 40 percent of army strength in World War I and has done so at times in Iraq; during the Korean War, deployed guard strength was almost 140,000. This ought not to conflict with the guard's traditional state missions or its evolving role in homeland defense, but the current Army National Guard end strength of less than 350,000 is probably inadequate, even if the active-duty force expands as suggested. A ten-year expansion program should strive to expand the Army National Guard to somewhere between 450,000 and 500,000 soldiers. Overall, a larger land force not only would be better able to sustain long-duration stability or constabulary missions, but could more rapidly prepare to engage in large-scale land campaigns.

- The requirement for building new partner land-force capacity is related to a continuing theme of this study: the need to bolster the institutional base of the army and Marine Corps. Military leaders increasingly acknowledge the ratio of combat units to support units and to overall service strengths is too high; contrary to received wisdom, the long-term problem is not the amount of "tooth" in U.S. land forces but the lack of a sustaining support and institutional "tail." The Bush administration's plan of expansion may begin to address the need, but is far from an adequate response. For example, the Gates Plan for the army increases active end strength to 547,000 in forty-eight BCTs, or one brigade for every 11,400 soldiers. A ratio better suited to emerging sustainment needs and the health of the services as institutions is probably closer to 13,000 soldiers per maneuver brigade. Thus an active force of at least sixty BCTs requires a total strength of about 800,000. An active Marine strength of 200,000 or more is likewise needed to flesh out the corps.

Structuring the Force

The process of defense transformation has had a profound effect on land-force structures, that is, on the institutional base of both the army and Marine Corps and on army unit structures. With the exception of the creation of a special operations command, marine unit structures remain largely unchanged. By contrast, the army's plan of modularization, unveiled by former chief of staff General Peter Schoomaker in 2004 and now largely complete in the active force, was described by the Government Accountability Office as "the largest Army reorganization in 50 years . . . Restructuring these units is a major undertaking . . . The Army's new units are designed, equipped and staffed differently than the units they replace."[4] In sum, the army began to remake itself just as its primary mission was changing—from rapid deployment and decisive combat to long-duration counterinsurgency and stability operations—and long before the nature of these changes was well understood. It is far from clear that the modularization process is optimal or even appropriate for the army's current and likely future tasks, and it may actually prove a handicap in fighting the Long War.

Army modularization was a response to the failures of the 1999 Kosovo campaign, and in particular the troubles of deployment faced by what was known as Task Force Hawk. The combat phase of the campaign lasted only seventy-eight days; it might have been concluded even more rapidly but for the political complications that limited operations, but it was at any rate complete before the army's task force, centered on AH-64 Apache attack helicopters but complemented by a heavy battalion for local protection, could participate. The subsequent stabilization campaign, though requiring a U.S. force of just two brigades, was disproportionately disruptive to the army. Overall, the service's performance raised a basic question about its relevance in the post–Cold War era.

Thus the army's primary concern prior to September 11 was improving its strategic deployability, and this concern became the lens through which the service considered the defense transformation process. Gen. Eric Shinseki, who had been the head of U.S. European Command during the Kosovo effort and later became chief of staff of the army, told Congress, "Our goal is to be able to deploy a combat brigade anywhere in the world 96 hours after receipt of an order to execute liftoff, a division within 120 hours, and five divisions within 30 days."[5] The army set about a ruthless process of trimming itself and its units to meet these benchmarks; the solution was "modularization," which emphasized the brigade as the basic tactical unit, trimmed the size of maneuver brigades from about 5,000 to 3,500, and reduced the size, structure, and emphasis on divisions, which had been the dominant element throughout the Cold War and through the 1991 Gulf War.

Modularization also was premised on the idea that a more transparent battlefield would allow for certain tactical economies of force structure. That is, the ability to see the enemy and understand the terrain—to lift the fog of war—would be so transformed by information and other technologies that land force units could be smaller yet equally effective in combat. Perhaps most notable was the reduction of maneuver brigades from three battalions to two: this seemingly innocuous change genuinely represented a revolutionary change in army thinking. Interestingly, the lone exception to this three-to-two reduction is the army's six Stryker brigades. Army leaders were worried about the survivability of the highly mobile but lightly armored wheeled Stryker and decided to retain a more robust structure in Stryker brigades.

At the same time, the modularization redesign was notable for what it did not do: it did nothing to redress the shortage of sustainment and support capabilities in the active army. During the Cold War, the army wanted to create as many immediately ready and capable combat units as possible in the active force, and so it placed many of the capabilities and units it needed for sustained operations in the reserve component. This was quite rational, given the anticipated nature of a conflict with the Soviets in central Europe, but the consequences were that, in stability or constabulary operations, the capacities of the active force, measured by such "low density" units as military police or civil affairs, were quickly consumed. This imbalance was exposed as early as the 1989 invasion of Panama, but persisted through the 2003 invasion of Iraq. The modularization reorganization created two new combat support and combat service support brigades, known as "sustainment" and "maneuver enhancement" brigades. Still, the basic active-reserve imbalance remains: for example, the army plans to have three maneuver enhancement brigades in the army reserve, fully ten in the Army National Guard, yet only three in the active-duty force. This is in part why the missions in Iraq and Afghanistan have required a continuous level of army reserve component mobilization of 80,000 to 100,000.

Thus, modularization has done little to improve the army's ability to conduct long-term stabilization campaigns. Despite a decade's experience in post–Cold War stability operations and the emerging requirements for long-term engagement in Iraq and Afghanistan, modularization reflected the priorities of the past. As army colonel Brian G. Watson has written:

> [Service] cultural aversion trumped experiential learning, and the Army embarked on a modernization path defined by a new operational framework—Rapid Decisive Operations . . . Its central operational framework—effects-based operations—integrated the application of precision engagement, information operations, theater enablers and dominant maneuver to produce a relentless series of multidimensional raids, strikes, and ground assaults throughout the battlespace . . . The Army, however, was not designed for the full task at hand. While the Army had perfected its ability to defeat any adversary swiftly, it also had mortgaged its ability to conduct protracted stability operations and deliver the enduring results the national strategy now insisted it achieve.[6]

Indeed, as our experience in Iraq reveals, the very successes of swift regime-change operations can, if all the organs of state security are destroyed or collapse along with the field forces, provide the tinder for insurgency. Stabilization operations are the other half—the truly decisive and important half—of "Rapid Decisive Operations."

The lessons of Iraq and Afghanistan are also, very clearly, that, in a modern insurgency, the distinction between combat and stabilization operations tends to vanish. Therefore, "handing over" from combat to "stabilization" units is impossible; brigade-level units have to retain, or be augmented with, the capability to conduct basic area security and stability operations, securing the local populace by reinforcing or replacing police and quasi-judicial functions and caring for or recreating the material infrastructure. And indeed, pushing these capabilities down to the lowest possible tactical echelon is proving the most successful tactic.

Creating an army capable of extended operations—both combat and stability—demands a force with more-or-less equally balanced elements: a sturdy service is like a sturdy, three-legged stool. Foremost among these elements, of course, is the frontline force of soldiers in the maneuver brigades and in their immediate division commands. This is often known as the "table of organization and equipment," or TOE army, but for the purposes of this study we will call it the field force. The second leg of a sturdy army is reflected in the so-called echelon above division, or EAD, support units at the corps and theater level; these units will be called the support force. The final crucial leg is known as the "table of distribution and allowances," or TDA army, and includes those soldiers in administrative units and what the army calls the TTHS accounts—for trainees, transients, holdees, and students. This is the institutional base. At the moment, of the 510,000 soldiers in the active army, about 180,000 make up the field force, 150,000 the support force, and 180,000 the institutional base. Indeed, the requirements of the Long War are such that this final leg, which reflects the institutional health of the service and its commitment to training and education, might need to be the strongest of the three. Yet the Pentagon's mania for management efficiencies and the army's desire to maximize its combat and deployment capabilities have meant that the institutional base has been gutted over a period of fifteen years.

But let us consider first the field force, the TOE army. The modularization process has significantly increased the number of maneuver brigades in the active army, from thirty-three old-style brigades to forty-two and soon forty-eight BCTs. Further, the service's modernization program, which is centered on suffusing improved information systems and command and control capacity throughout the force, is well on the way to making each headquarters far more capable. But as discussed above, these improved headquarters have fewer regularly assigned subordinate units. In particular, the thirty-six heavy and light BCT designs have only two subordinate maneuver battalions, compared to three battalions in the previous structure; the six Stryker BCTs retain the three-battalion formation. The design of the six BCTs to be added under the Bush administration's expansion plan is not yet determined and depends on the cost of equipment.

Thus the first goal for a serious program of army expansion should be to restore the third maneuver battalion to the BCT design. Adding a third battalion to the heavy and light infantry BCTs—a manpower bill of about 600 each—would restore tactical flexibility and robustness. Restoring the third battalion to the currently planned active force of forty-eight BCTs would thus total about 30,000 additional soldiers; adding the twelve new BCTs as described above would call for a further 60,000 soldiers. Equally, the active army force structure needs additional maneuver enhancement brigades to augment combat units and provide the engineering, military police, ordnance disposal, civil affairs, chemical, and other capabilities that are so essential in extended stability operations. In addition to providing enhanced capacity to maneuver brigades moreover, these support brigades can provide the headquarters and command functions for more complex and advanced stability efforts, such as in more pacified areas where the balance between combat and reconstruction operations has shifted. The army's current ratio of maneuver enhancement brigades to BCTs—three to forty-two (and about to become three to forty-eight under the Bush administration's plan)—reflects the service's hollow commitment to sustained stability operations without immediate and heavy resort to reserve component forces. This study recommends that, commensurate with the increase in the number of BCTs to a total of sixty as described above, the active army should include fourteen maneuver enhancement brigades; a preferable ratio would be to have one for each division-level headquarters. At an average of about 2,500— the size can vary depending on the number of subordinate units assigned—the increase would account for about 30,000 additional soldiers. In sum, the total number of soldiers in the "field force" would increase by about 120,000, from the current 180,000 to approximately 300,000.

The second priority is to restore the institutional base of the service. Supporting a field force of three hundred thousand will demand an institutional base of like size. Of this, the largest proportion will be reflected in the TTHS personnel accounts described above; in particular, the number of soldiers in basic enlisted and officer training and the number of NCOs and officers in higher levels of training and education will increase significantly. A healthy service should expect to have 15 percent of its total strength in these accounts (as compared to the current 12 percent), which would mean about 125,000 from a force of 800,000. Army agencies such as Training and Doctrine Command, which develops service tactics and operational concepts and operates the army's training facilities, should expect to grow as well; TRADOC currently accounts for less than 45,000 soldiers out of 510,000 and could increase to roughly 70,000 in an 800,000-soldier force. Altogether, the size of the institutional army might rise to 275,000.

The third leg of the stool, the higher-level support force, is harder to approximate. But even an army of 800,000 would suffer from constraints; greater risk would have to be assumed in the support force of units assigned above the division echelon. The increased end strength recommended in this report ameliorates the problems of an "unbalanced" army structure but does not eliminate them altogether. Higher-echelon units perform important "enabling" functions to divisional units and often provide the tools with which senior commanders guide operations or reinforce their main efforts.

While ideally this element would be roughly equal in size to the field force and the institutional base, some of the risk would be mitigated by increasing the support capabilities within the divisions through the proliferation of maneuver enhancement brigades. Similarly, some of these units, such as corps-level artillery support, may be less in demand. And finally, it is reasonable to expect that the army's reserve components can continue to provide some of these capabilities, while understanding that the greatly enlarged active force can lighten the burdens of too-frequent deployments. The total strength of this element of the army might grow to 225,000.

Equipping the Force

While most public attention has focused on the personnel demands of Iraq and Afghanistan (and to a far lesser degree on the requirements for the Long War), much less debate and discussion have been devoted to the equipment needs of U.S. land forces. Indeed, the few issues that have achieved any prominence, such as the question of individual body armor and armor plating for high mobility multi-purpose wheeled vehicles (HMMWVs), the utility of mine resistant ambush protected (MRAP) vehicles, and the defeat of improvised explosive devices, have obscured rather than clarified matters. The Bush administration was and remains highly reluctant to increase investment in land forces, a legacy of the transformation movement and the belief that U.S. forces would soon withdraw from Iraq and Afghanistan, and captured by former defense secretary Donald Rumsfeld's famous comment, "You have to go to war with the Army you have, not the Army you want." For their part, congressional Democrats have preferred to exploit the issue politically more than substantively, as is suggested by a July 2007 press release by Sen. Joseph Biden, titled "Sen. Biden Again Demands Investigation into MRAP Delay; 'Pentagon Needs to Come Clean'":

> Those on the frontlines knew that they needed better protection against the road-side bombs that were killing their comrades; they knew we had the technology—but

their requests were repeatedly ignored by the Pentagon
and by a president who has claimed all along that he lis-
tens first and foremost to those in the field.[8]

But the first fact that shapes the equipment needs of U.S. land
forces is simply the need to refit, repair, and replace what has been
worn out or destroyed by operations in Iraq, Afghanistan, and elsewhere. In short, the services' major equipment stocks—that is, the
investments of the Reagan buildup of the late Cold War years—are
severely depleted, even allowing for the large drawdown of the 1990s.
Two recent studies on the state of army and Marine Corps equipment
came to similar conclusions. The fleet of army tanks, Bradley infantry
fighting vehicles, and helicopters in Iraq continues to be used at about
five to six times the programmed tempo; trucks and other support
vehicles are being used at *ten times* the normal rate. Marine usage rates
are only slightly lower.[9] The essential quality of the original systems
and historically high levels of equipment readiness mean that the
problem is not the destruction of vehicles but their service life; as
marine Lt. Gen. John Sattler remarked, "If we bought something to
last for 21 years, I'll be honest; I think we'll get three years out of it."[10]
But the repair bills have been high. By the end of 2007, the army was
estimating that it would have repaired a total of 557 aircraft, seven-
teen hundred tracked vehicles, more than eight thousand HMMWVs,
eighteen hundred trucks, twelve hundred trailers, almost forty thousand small arms and about seventy-four hundred generators. The cost
of refitting the previously existing army and Marine Corps has
exceeded $30 billion through 2007, and it appears that the overall
level and quality of prepositioned stocks may have eroded.[11]
Regardless of any likely improvement in Iraq and Afghanistan, there
is no reason to expect the situation to significantly improve; the
equipment costs are a measure of force size, the age of the systems,
the harshness of the environment, and operational tempo.

A second fact shaping U.S. equipment needs is that the duration
of current conflicts and the nature of the combat have imposed new
equipment requirements; U.S. land forces must adapt technologi-
cally as well as tactically and doctrinally to unforeseen realities. To

take a signal example: the introduction and development of the improvised explosive device in Iraq has made a fundamental change in the challenges faced by soldiers and marines; they are fighting on a genuinely transformed battlefield. An excellent study by Andrew Krepinevich and Dakota Wood of the Center for Strategic and Budgetary Assessments succinctly captures the devastating effect of IEDs:

> For the period March 2003 to early August 2007, 1,496 of a total 3,037 deaths due to hostile causes were attributed to IEDs (49.5 percent). From January 2005 to early August 2007, the percentage increased to 65 percent. And from March 2007 onward, the percentage of hostile deaths attributed to IED attacks continued to rise to 72 percent. Stated another way, for the conflict as a whole, from March 2003 to August 2007, IEDs have accounted for half of all deaths due to hostile causes.[12]

These numbers taken in the abstract probably underestimate the strategic effect of IEDs: the combination of casualty-averse elite and military opinion, growing unease with the Bush administration's conduct of the war, and political pressure from opposition politicians (reflected in the quotation from Sen. Biden above) has been an important factor shaping public dissatisfaction with the Iraq mission. Despite the impolitic comments of Secretary Rumsfeld, the Defense Department has been highly sensitive to the IED threat. It established a Joint IED Defeat Organization, with a budget of $12 billion from 2006 through 2008 alone. But its efforts, along with constant tactical innovations by soldiers and marines on the ground, are hardly the whole story; indeed, in a political and strategic sense, they pale in contrast to the controversy over individual body armor, additional armor for HMMWVs, and the MRAP program. Even though crash programs have resulted in improved body armor and the upgrading of the roughly twenty-one thousand HMMWVs in Iraq, Rumsfeld's replacement, Robert Gates, authorized the purchase of up to seventy-eight hundred MRAPs in May 2007, at a cost of nearly $8 billion.

"The MRAP program should be considered the highest priority Defense Department acquisition program," he said.[13]

The Pentagon's hot-and-cold approach to the MRAP and to other, less expensive war-related investments (the MRAPs cost about $800,000 each, or eight times an up-armored HMMWV) represents not just a failure to respond to battlefield imperatives but a more profound uncertainty about how to think about equipment modernization. The current requirement for seventy-eight hundred MRAPs is many times larger than the original requirement, but still less than half the total proposed by advocates who would like to substitute the MRAP for the HMMWV entirely. And while there is probably a long-term mission for a unique vehicle for convoy escorting, patrolling, and clearing roads, understanding the broader requirement in environments other than Iraq—the MRAP is far less useful in Afghanistan, for example—has merit. Further, given that dismounted operations are essential to counterinsurgency and stability operations, the impulse to "get out and walk," to quote Lt. Gen. Raymond Odierno, commander of coalition forces in Baghdad, is essential. "Vehicles like the up-armored HMMWV limit our situational awareness and insulate us from the Iraqi people we intend to serve."[14] The MRAP is even more isolating.

If differentiating wartime expedients from long-term investments has been a source of confusion in the direction of American land forces, that is in part because the current modernization programs of both the army and Marine Corps were conceived prior to the attacks of September 11. Further, those programs that did not originate in the late Cold War years were products of the transformation movement and were justified as improvements in expeditionary capability and, in particular, as meeting the need for rapid, strategic deployments—a need that has lower priority in our current Long War. Finally, the army's Future Combat System, by far the most ambitious and expensive land force modernization effort in a generation, is a profoundly misunderstood program (though a necessary one, as will be argued).

But consider first the state of Marine Corps equipment modernization. The marines' from-the-sea stance imposes a wide variety of

technical and tactical challenges, as well as the bureaucratic challenges that stem from being part of the Department of the Navy. To
begin with, the marines need ships from which to operate, and navy
ships and submarines in escort. Setting aside the destroyers and submarines that provide protection and fire support, the marines deploy
in three basic kinds of ships: an aircraft-carrier-like amphibious
assault ship designated an LHD, which carries the large complement
of helicopters needed to ferry marines ashore and a flight of eight
AV-8B Harrier jump-jets; an amphibious transport dock ship,
known as an LPD, which carries much of the combat support and
combat service support needed for both marine air and ground
units; and a dock landing ship, an LSD, a cargo ship with somewhat
different characteristics. In a simple but profound way, the state of
marine shipbuilding programs is inseparable from the state of the
larger naval shipbuilding industry. While the current marine ship
inventory is relatively modern (though the marines still operate from
a good number of older-generation ships), the future is not bright,
given the steadily increasing costs of shipbuilding, excess industrial
capacity, and shrinking shipbuilding budgets. Navy Secretary
Donald Winter recently announced further trimming of budgets and
programs. "A steady course has been set, but there are storm clouds
on the horizon and we need to alter course to stay clear. We need to
arrive at better alignment between Navy and industry," he says.[15]
What that really means is that the navy cannot support its
announced shipbuilding program or goals.

It's the other aspects of marine equipment—those related to the
corps' ability to project power ashore—that are truly worry-making.
The marines fly old aircraft and operate too much obsolescent
ground equipment. The bright side for the marines is their fleet of F-
18 fighters, but these can operate only from large-deck navy carriers
or from large airfields ashore. The AV-8B Harrier was a tremendous
improvement on the original in almost every capacity and has
become an integral element in marine fire support operations, but
that upgrade program was, again, a product of the Reagan-era investments, and the fleet has aged. "The Long War on Terror has resulted
in aircraft use rates far greater than designed or programmed on

Marine Corps aircraft," Gen. James Conway, commandant of the Marine Corps, told the Senate Armed Services Committee. "All USMC aircraft are operating at two to four times their programmed rate."[16] The replacement for both the F-18 and, more pressingly, the Harrier, is the F-35 Lightning Joint Strike Fighter, an immense and international program worth hundreds of billions of dollars. Arguably, the marines, along with those international partners like Great Britain that face a similar problem of aging Harriers, need the F-35 more desperately and immediately than does the U.S. Navy or the U.S. Air Force. Says Conway: "F/A-18D . . . aircraft will reach the end of their service lives before replacement aircraft become available. These shortfalls underscore the urgency for the F-35B program to remain on schedule."[17] As with shipbuilding, however, the marines' fate depends on many factors outside the corps' control.

But the real nightmare for the marines is the state of their rotary-wing aircraft fleet. Increasingly, helicopters are the key to the marines' continued relevance as a sea-based force. But their inventory of CH-46 and CH-53 helicopters is critically old; the fleet is on the verge of collapse.

> The average age of the Vietnam-era CH-46E was over 35 years old when the U.S. invaded Iraq . . . The situation with the CH-53E Super Stallion is similar. Currently the Corps has only 150 of the 160 it needs and starting in 2010 it may have to take as many as 15 out of service every year.[18]

The marines have been working on a replacement, the "tilt-rotor" V-22 Osprey, for a generation. Constant technical problems and program turbulence—the V-22 project was one of the first victims of the post–Cold War era, terminated by then secretary of defense Dick Cheney in 1992 and revived by the Congress—have combined to leave the Marine Corps dangerously short of helicopter lift at a time when the operational requirements have skyrocketed. Solving this equipment shortfall is the most crucial issue for the future of the Marine Corps.

Finally, the state of the marines' ground combat equipment is likewise worrisome. There are a number of bright spots, in particular the Light Armored Vehicle, a highly mobile eight-wheeled vehicle generally like the army's Stryker. The marines have few heavy tanks and employ the M1, as does the army, and the marines' howitzers remain quite serviceable. The most problematic piece of marine equipment is the AA-7A1 amphibious assault vehicle, first introduced in 1972. The huge vehicle is very slow in the water, making only five knots, and very vulnerable and lightly armed on land. The replacement, the so-called Expeditionary Fighting Vehicle, moves much faster at sea, skimming above the water rather than ploughing through it, and includes a 25mm cannon like the army's Bradley fighting vehicle. It is supposed to be introduced beginning in 2015, but budgetary and development woes make that date doubtful.

In sum, the proper course of marine modernization is inextricably linked to the role to be played by the corps in the Long War. If the marines return to their sea-based mission—one perhaps vital to the more indirect strategies for the Long War—then the direction of investments is better defined and the need for heavier weaponry moderated; priority can and should be given to aviation and especially rotary-wing aviation. But if the marines must still be used as a substitute for a too-small army, their equipment problems can only get worse. The marines have been particularly vocal about the need for MRAPs, precisely because their vulnerabilities in sustained operations ashore have been so dramatically exposed. However, too many MRAPs make sea-based deployments—inherently difficult to begin with—an even greater challenge. The marines face a defining moment in their history, and their future course depends first and foremost on the course of the U.S. Army.

Modernizing the army in the midst of a long-running but constantly shifting conflict, while expanding it and reshaping its entire organizational structure, is a daunting challenge. There is a solution to the puzzle, the Future Combat System program, though unfortunately, widespread lack of understanding about the program is a dangerous point of failure. The army has unfortunately done little to help its own cause: it has failed to articulate a rationale for the

FCS, and has emphasized aspects of the program (its solution to the problem of strategic deployability) that are not crucial in the Long War (where the real test is not how fast we arrive but how long we stay). And it has further confused the discussion about modernization by getting sidetracked in a process debate about employing a "lead systems integrator"—the Boeing Corporation—to manage the vast undertaking. Many in Congress remain uncertain about the purpose and intimidated by the price of the FCS program, and as a result units in Iraq and Afghanistan find themselves improvising in order to supply the information and communications systems that are the backbone of the FCS concept.

The FCS project is the largest land-force modernization effort in almost forty years. The "Big Five" programs of the Reagan years (the M1 tank, the M2 Bradley fighting vehicle, the UH-60 Black Hawk and AH-64 Apache helicopters, and the Patriot interceptor missile) were in reality not five separate programs but a comprehensive equipment modernization effort based on the doctrine of "AirLand Battle," a concept seen as the operational solution to the threat posed by the massive Soviet forces stationed in Eastern Europe. In essence, the Reagan buildup financed a new army, with an unparalleled ability to conduct high-intensity ground maneuver campaigns. These were the capabilities that twice brought Saddam Hussein's army to its knees in such a decisive fashion, ultimately evicting the Iraqi regime in a three-week *blitzkrieg* in 2003. But when the Cold War ended, the army, like all the other services, found itself hamstrung by modernization efforts aimed at improving these capabilities at the margin, in response to Soviet improvements. Alone of the U.S. services, however, the army was actually forced to terminate major systems like the Crusader cannon and the Comanche scout helicopter. The army was made to go back to the drawing board.

What emerged was the FCS. At its heart, this is an effort to "network" an entire force—a genuinely transformational undertaking in the context of land combat, where the fog of war is thickest, especially in irregular warfare. It is also an attempt to answer the three most important but difficult questions posed by land commanders throughout history: *Where am I? Where are my friends?*

Where is the enemy? These timeless questions transcend questions about the size, armor protection, and firepower of individual systems; about whether they fit comfortably inside an air force airlifter; or about whether the army possesses the internal management know-how to run such an ambitious and cross-cutting modernization program. For all the hyperbole about "network-centric" warfare, the truth is that better combining and coordinating aspects of land power is critical to success on opaque battlefields. This is doubly true for a force that will remain relatively small in comparison to the size and diversity of missions required of it.

Alas, it is difficult to explain the concept of the FCS network without descending into the jargon of transformation or information systems; despite the growing presence of geek-speak in modern life, it can often sound as though the system is itself the object, rather than a means to greater military effectiveness. But the goal is a simple and important one: to make soldiers at the lowest echelon—the platoon, the squad, the section, even the individual soldier himself—exponentially more effective. And because the nature of modern irregular warfare restores the focus of war to these low levels, building the FCS network ought rightly to be the focus of army—and all U.S. land force—modernization. We need to make the dismounted soldier on patrol as knowledgeable, as lethal, as survivable, and as capable as possible.

The five-part FCS network effort broadly resembles the approach common to modern information systems. At the base is a core set of data standards that govern how the system will work. The second part builds a set of communications and other "pipelines" that can transport large and growing amounts of data securely and rapidly, ultimately delivering it to soldiers conducting operations. Third is what in an office environment is known as "middleware," such as communications and monitoring software and databases. Fourth are the particular applications—word processing, spreadsheet, or media programs are the civilian analog—that soldiers will actually use. And finally there are the sensors and platforms that generate and manage improved battlefield awareness, and make soldiers themselves the primary sensor in the network.[19]

The centrality of the network entirely inverts the Defense Department and army paradigms for equipping the force. Rather than constructing the array of dismounted infantry, mobile force, fire support, other combat support, and logistics support capabilities and then combining them, which is the traditional approach to "combined arms" operations, the FCS approach invests first in the links and then in the "plug and play" components. It is thus an approach intended to further amplify the value of current platforms, like the Abrams or Apache, that remain largely unchallenged in terms of their mobility, firepower, survivability, and other traditional measures of effectiveness. It also paves the path for fielding new systems, particularly unmanned systems both ground and air, that can be more readily tailored to immediate needs.

The FCS program has been structured to try to field elements of the system and a number of new capabilities as rapidly as possible. In 2008, the army intends to initiate the first of three "spin outs," testing prototypes at the newly created Army Evaluation Task Force at Fort Bliss, Texas, and then introducing improvements into the current force. The second spin out, scheduled for 2010, is to include an active protection system designed to intercept incoming rounds, such as rocket-propelled grenades, before they hit U.S. vehicles, and a mast-mounted suite of sensors for the Stryker vehicle. The third spin out, planned for 2012, may include the widespread introduction of robotic and unmanned vehicles, and indeed small "battlebot"-style prototypes are already being used effectively for building-clearing operations in Iraq. And quite sensibly, the army has deferred decisions about new generations of major land combat vehicles until later in the program.

While its post–Cold War efforts at modernization have suffered from a great degree of uncertainty and a great number of false starts, the army has lurched into a sensible approach (by contrast, the other services continue to spend vast and ever-increasing amounts on ever-reduced numbers of late Cold War systems whose value in future wars is very hard to assess). Further, it is an appropriately flexible approach to force modernization, one that maximizes the value of past investments and older systems while also maximizing options for the future. To repeat: the value of the network should be measured across the force from the individual soldier and the smallest unit to the most senior commands. It is the technological expression of the need discussed throughout this report for good judgment and decisive initiative at moments of inherent uncertainty.

Yet the army's modernization efforts and the FCS program do need to be more sharply honed if individual soldiers, particularly dismounted infantrymen, are to derive the greatest benefit from them. The past "Land Warrior" effort was scrapped in 2007 and renamed the "Future Force Warrior" program, but it remains to be seen whether a new name reflects new investment and new focus. While many aspects of information technology are adaptations of commercial products, the need to make them useful to soldiers in close combat in urban and other complex environments poses a challenge. It involves thinking of a soldier's weapon, helmet, uniform, and navigation and communications equipment in a systemic fashion, as well as considering the soldier as a node in a larger network; for example, there is an effort underway to allow soldiers in Stryker units to interface and operate systems on the vehicle after they have dismounted.

In sum, the primary goal of all land-force modernization, both army and marine, should be to exponentially increase the effectiveness of the individual, dismounted soldier. This is a need made all the more compelling by the requirements of extended irregular warfare. Tactical and strategic success depends on soldiers' abilities to interact with and protect diverse peoples; vulnerability and defeat are likely to be found in fear of or reaction to casualties. The remarkable performance of American land forces in Iraq and Afghanistan—where it can be fairly said that they have rescued American strategists from the worst consequences of their mistakes—is a reflection of superb training. If we are now to ask that soldiers supplement their tactical prowess with a strategic awareness that would come from a broader education, we ought also to give them all the advantages that technology can afford.

Paying for the Force

Looked at from one perspective—one focusing on the difference, as Secretary Rumsfeld might have said, between the forces we have and the forces we need—the costs of achieving the land power goals outlined above are steep. But looked at in the context of the American economy, they are affordable, even modest. Simply increasing overall defense spending by one percentage point of gross domestic product, a penny on the dollar, ought to provide sufficient resources.

To be sure, the cost is impossible to predict with green-eyeshade precision. The course of any war is hard to predict; as is ruefully said, the enemy gets a vote. Nonetheless, it is possible to project a rough range of estimates. The true test is to resist the temptation to assume away difficulties, to avoid the embrace of "Rosy Scenario."

In this regard, it is important to understand the costs of fighting the Long War. The total Pentagon spending on Long War–related activities from September 11, 2001, through the 2008 fiscal year— and this depends on whether Congress approves the supplemental defense appropriations more or less as requested by the Bush administration—will be approximately $660 billion.[20] It is important to remember that these are simply the additional costs over and above the "baseline" defense budget; that is the cost of expanding, by calling up reservists, and operating the force that previously existed.

Even a cursory glance at President Bush's requested $190 billion budget amendment for 2008 indicates why, as a crude benchmark, many of these expenses will recur in the coming years. Of that total, about $77 billion, by far the largest share, is consumed by what the Pentagon calls "operations and maintenance," which includes not only the costs of logistics support but the pay and benefits of mobilized reservists. In other words, this is a reflection of the size of the total committed force—about 320,000 military personnel will rotate through the theaters of war in 2008—and the pace of operations. The next largest slice, almost $47 billion, goes to reconstitute the force: equipment replacements, repairs, replenishment of munitions, and some upgrades. Again, given the size of the force and pace of operations, this cost could remain relatively constant, especially given that the amount due to wear and tear, as opposed to actual combat losses, represents the bulk of the total. The third big chunk of the costs come under the $31 billion category of "force protection," including personal body armor, armor kits for thin-skinned HMMWVs and trucks, and new procurements like the MRAP, which accounts for almost $18 billion alone. The remainder of the total is equally likely to be recurring, if not escalating, costs: U.S. military intelligence, anti-IED devices, support to the Iraqi and Afghan armies, and so forth. In a crude sense, one can estimate the annual ongoing cost of the war at about $150 billion; the costs in 2007 and 2008 may well be above a longer-term average, reflecting the fact that the Bush administration was slow to adjust to the many realities of the Long War. This estimate is still less than 1.1 percent of total U.S. GDP.

But this report has argued throughout that it is the longer-term investments in the size, structure, equipment, and institutional base of U.S. land forces that will provide the telling strategic difference and pave a road to victory in the Long War. And indeed, despite the many wartime controversies, there has been broad bipartisan support for meeting the budgetary requirements of the force in the field. The test of political leadership will be paying the costs to meet less immediately apparent needs. And, to underscore another theme of this report, the hardest question on the test is how to address the needs of the army, which is the force bearing the brunt of the Long War. The needs of the Marine Corps are relatively less: sufficient funding to flesh out and modernize the force as it already stands, not a radical expansion or reorientation.

Of the approximately $480 billion baseline defense budget requested by the Bush administration for 2008, about $128 billion, or between 26 and 27 percent, will be spent for the "baseline" Army. That is a modest improvement on the levels of 2001, when the Army received 25 percent of defense spending, but hardly enough, in gross terms or in proportion, to build the larger, more capable force that is needed. Further, the 2008 budget shortchanges the needs of the current army by about $10 billion.[21] In rough terms, the cost of the current army, raised, trained, and equipped according to its doctrine, is about $275 million per thousand soldiers. And it is the general conclusion of this report that even this figure underestimates the costs of the expansion, modernization, and restructuring that is required; for purposes of estimating the cost of the army we need, we should expect to pay about $300 million per thousand soldiers.

Thus, the total cost of an army structured around an eight hundred thousand–strong active-duty force would be about $240 billion. That seems like a lot of money. But in the context of the overall U.S. economy, the sticker shock diminishes. The United States will produce about $14 trillion worth of goods and services in 2008. The Congressional Budget Office forecasts that by 2017—about the year by which, this report estimates, such a buildup would be complete—U.S. GDP will exceed $21 trillion.[22] Thus, the total burden of an expanded army, by the time the process neared completion, *would remain at less than 1.2 percent of projected GDP.* So even if other increased defense expenditures were required, and even if the costs of the war continued at about 1 percent of GDP, the program described in this report is eminently affordable.

Conclusion

The strategic pause of the 1990s proved a mirage, but the American military—its policy, structure, size, and funding—has yet to catch up to reality. Our military today is not configured to address the counterterrorism and counterinsurgency missions required of it, now and for the foreseeable future, nor to sustain a large-scale presence in post-conflict or stabilization scenarios. That is why the effort to maintain forces in Iraq and Afghanistan is proving to be such a strain.

Today's wars are being fought with armed forces designed in the 1980s to excel in a different kind of combat—short-term, high-intensity combat that was expected to lead to rapid and complete victory or defeat in one major theater. Priority was given to getting soldiers and tanks into the fight quickly in the belief that support elements, headquarters, and reinforcements could follow more slowly. But this priority is out of sync with today's needs and has created an imbalanced active-duty force that faces grave challenges in sustaining long-term deployments and carrying out its varied, numerous missions.

In fact this priority was always out of sync with the true needs of U.S. land forces. The focus on "rapid decisive operations" at the expense of the more habitual operations of land forces is an anomaly in American military policy. It is not just current conflicts that require a long-term deployment of relatively large numbers of American forces; every successful major conflict since 1945—Germany, Japan, South Korea, Panama, Iraq (1991 and 2003), Bosnia, Kosovo, and Afghanistan—has required the same thing. By the same token, the rapid withdrawal of American forces after conflict has generally signaled or led to failure—Vietnam, Lebanon in the 1980s, Somalia, Haiti. Long-term postconflict deployments are not an innovation of the Bush administration; they are the way America wins wars, and they will continue to be the coin of the realm in any effort to achieve U.S. interests that requires the use of significant military force.

It is time to return the military to a traditional understanding of the challenges it faces and the likely solutions to them. That means above all addressing the needs of the ground forces. This is not to say that maintaining high-quality, modern, ready, and numerous air, sea, and space forces should not be a priority for the United States. Those forces, dominant today, will face potential challenges in the future for which they must be prepared. But the ground forces require immediate help to meet the challenges of the moment and to prepare for the challenges they may face tomorrow. The strain the current struggles in Iraq and Afghanistan are imposing on the ground forces makes clear the need for change; these are at the low end of the spectrum of conflict and suggest that potential future crises at the higher end of the spectrum will prove more difficult still. Involvement in Nigeria, the Congo, and Pakistan, for instance, to say nothing of Iran, might all require even greater troop commitments. Currently, the deployment of NATO forces in Afghanistan has kept that burden relatively low for the U.S. military (only three combat brigades out of seven or more deployed in the country are American). But smaller-scale crises in Somalia, Darfur, or elsewhere for which international forces were not available could well pose a significantly larger challenge for U.S. ground forces.

The strain on the ground forces, which originated in the crises in the Balkans in the 1990s, has forced them to accept increasing imbalance among components and missions. Rebalancing these forces—and providing them the capabilities they need to undertake long-term postconflict and stabilization missions without compromising their ability to fight and win major wars and without accepting undue strain—is an urgent priority. The ground forces must begin by increasing their numbers dramatically, though greater numbers are only a starting point. Growth in the ground forces also requires that they be reshaped, restructured, and reequipped. More than these, it requires that forces be distributed correctly across services and service components—that the combat forces and the institutional base, the army and marines, and active duty and reserve forces be rebalanced. Getting these balances right would ease the burden on soldiers, who would need to be deployed less often, and on marines and members of the National Guard, who could pursue their true functions instead of being called on to make up any shortages in army manpower. It would also improve the army's ability to train and educate soldiers most effectively.

These capabilities, in turn, would enable the army to become a truly full-spectrum force, ready to handle not only high-end conventional warfare, but also counterinsurgency, foreign internal defense, and many other missions at the lower end of the spectrum of conflict. In particular, they would allow the U.S. military to undertake the sort of operations it is likely to face in the coming years, without subjecting the ground forces to unacceptable strain. Even the increased, rebalanced force we propose, it should be understood, would accept considerable risk. It would not be adequate to contend with such operations on the high end (Pakistan or Iran, for example) except under extreme strain (our current forces, of course, could not reasonably hope to undertake such high-end operations short of full-scale mobilization). The expansion of the ground forces proposed in this report, then, is really a floor—that is, the minimum configuration of American land power necessary to mitigate the unacceptable strain and intolerable risk the current force accepts.

Certainly the cost of building the necessary land forces is very large. The cost of failing to do so is incalculably larger. Land power is the key to success in the Long War, and the importance of succeeding in that fight, for our own country and the world as a whole, can hardly be exaggerated. The political future of the Islamic world holds the key to future peace and prosperity. The region is immeasurably important to the international economy, particularly to developing great powers like China and India, and to international security. The opportunity costs to be incurred by the collapse of an American endeavor to guarantee the security, stability, and integration of the Muslim world into the political order of the twenty-first century are tremendous. More profoundly, the cost in lives of such a collapse is almost too horrible to contemplate.

Notes

Chapter 1: The Mission

1. Frederick W. Kagan, *Finding the Target: The Transformation of American Military Policy* (New York: Encounter Books, 2007), 243.

2. Donald Kagan and Frederick W. Kagan, *While America Sleeps: Self-Delusion, Military Weakness, and the Threat to Peace Today* (New York: St. Martin's Press, 2000), 429–30.

3. Department of Defense, "Report of the Secretary of the Army," in *Annual Defense Report* (Washington, DC: Government Printing Office, 1995), http://www.dod.mi/execsec/adr95/army_5.html.

4. The White House, "A National Security Strategy of Engagement and Enlargement" (Washington, DC: February 1995), http://www.au.af.mil/au/awc/awcgate/nss/nss-95.pdf.

5. F. Kagan, *Finding the Target*, 245–51.

6. See D. Kagan and F. Kagan, *While America Sleeps*, 296–99.

7. Department of Defense, *Quadrennial Defense Review Report*, September 30, 2001, http://www.defenselink.mil/pubs/pdfs/qdr2001.pdf.

8. See Caspar W. Weinberger, "The Uses of Military Power," remarks prepared for delivery to the National Press Club, Washington, DC, November 28, 1984, http://www.pbs.org/wgbh/pages/frontline/shows/military/force/weinberger.html; and Colin Powell, "U.S. Forces: Challenges Ahead," *Foreign Affairs* 71, no. 5 (Winter 1992): 32–45.

9. The military is a large and complex organization, and so this generalization has limits. Army doctrine and the public statements of military leaders repeatedly asserted the importance of peacekeeping and similar operations other than war, and some organizations within the military did good work in thinking about such operations. But the intellectual energy of military leaders was not generally focused on these operations, and very little was done in reality to support them, despite a decade of nearly continuous requirements for them.

10. The classic work on American Cold War grand strategy is John Lewis Gaddis, *Strategies of Containment: A Critical Appraisal of American*

National Security During the Cold War (New York: Oxford University Press, 2005).

11. See D. Kagan and F. Kagan, *While America Sleeps*, 237, and F. Kagan, *Finding the Target*, 7, for references and details.

12. See Mary Habeck, *Knowing the Enemy: Jihadist Ideology and the War on Terror* (New Haven, CT: Yale University Press, 2007), for an introduction to al Qaeda's ideology. See Frederick W. Kagan, "The New Bolsheviks," *National Security Outlook*, November 2005, American Enterprise Institute, http://www.aei.org/publications/pubID.23460/pub_detail.asp; and Frederick W. Kagan, "Al Qaeda in Iraq," *The Weekly Standard*, September 10, 2007, 22–33, for discussions of al Qaeda's application of its ideology both globally and in Iraq.

13. A key indicator of this shift was the explicit rejection of "threat-based" planning and the adoption of "capabilities-based" planning that pervaded the 1990s defense debate and was seized upon by the Bush administration when it came to power in 2001.

14. For a detailed discussion of Iran's involvement in the Iraq War via the Jaysh al Mahdi and other forces, see Kimberly Kagan, "Iraq Report #6: Iran's Proxy War against the United States and the Iraqi Government," August 29, 2007, http://www.understandingwar.org/IraqReport/IraqReport06.pdf.

15. F. Kagan, "The New Bolsheviks."

16. Ibid.

17. The flow of weapons and advisors, and the complex networks Iranian agents have established in Iraq, have been revealed in a number of press conferences and press releases by officials of the Multi-National Force—Iraq, and also by the U.S. Embassy in Baghdad. These releases are summarized in K. Kagan, "Iraq Report #6: Iran's Proxy War."

18. See Robin Wright, "Iranian Flow of Weapons Increasing, Officials Say," *Washington Post*, June 3, 2007, http://www.washingtonpost.com/wp-dyn/content/article/2007/06/02/ AR2007060201020.html.

19. See Christine Wormuth, *The Future of the National Guard and Reserves: Beyond the Goldwater-Nichols Phase III Report*, Center for Strategic and International Studies, July 2006, http://www.csis.org/media/csis/pubs/bgn_ph3_report.pdf.

Chapter 2: What Kind of War?

1. John Lewis Gaddis, *Strategies of Containment: A Critical Appraisal of American National Security During the Cold War* (New York: Oxford University Press, 2005), 162–65.

2. See Department of Defense, *Quadrennial Defense Review Report*, September 30, 2001, http://www.defenselink.mil/pubs/pdfs/qdr2001.pdf; and Department of Defense, *Quadrennial Defense Review Report*, February 6, 2006, http://www.defenselink.mil/ qdr/report/Report20060203.pdf.

3. T. X. Hammes, *The Sling and the Stone* (St. Paul, MN: Zenith Press, 2006); see also William Lind's various articles on "fourth-generation warfare," e.g., "The Changing Face of War: Into the Fourth Generation," *Marine Corps Gazette*, October 1989, 22–26.

4. See for example Hammes, *The Sling and the Stone*, and Lind, "The Changing Face of War."

5. These arguments are, naturally, most common in the navy and air force communities.

6. See Frederick W. Kagan, *Finding the Target: The Transformation of American Military Policy* (New York: Encounter Books, 2007) for a review of competing visions of warfare in the 1990s and references.

7. See Bob Woodward, *Bush at War* (New York: Simon & Schuster, 2003), 43.

8. See Charles E. Heller and William A. Stofft, eds., *America's First Battles, 1776–1965* (Lawrence, KS: University Press of Kansas, 1986), for an excellent treatment of America's failure to prepare for future conflict and the consequences of that repeated failure.

9. Frederick W. Kagan, *No Middle Way: The Challenge of Exit Strategies from Iraq* (Washington, DC: American Enterprise Institute, 2007), http://www.aei.org/publications/pubID.26760/pub_details.asp.

10. Ibid.

11. Donald Kagan and Frederick W. Kagan, *While America Sleeps: Self-Delusion, Military Weakness, and the Threat to Peace Today* (New York: St. Martin's Press, 2000), 299.

12. Gaddis, *Strategies of Containment*, 162–65

13. F. Kagan, *Finding the Target*, 369–71.

14. Authors' conversations with senior French officers, June 2006.

15. See Hammes, *The Sling and the Stone*, and Lind, "The Changing Face of War."

16. See Frederick W. Kagan, "Al Qaeda in Iraq," *The Weekly Standard*, September 10, 2007, 22–33, and Frederick W. Kagan, "Why We're Winning in Iraq: Anbar's Citizens Needed Protection before They Would Give Their 'Hearts and Minds,'" *Wall Street Journal*, September 28, 2007.

Chapter 3: Case Studies

1. See Caleb Baker, Thomas Donnelly, and Margaret Roth, *Operation Just Cause: The Storming of Panama* (New York: Lexington Books, 1991).

2. Quoted in Lance Morrow, "The Peacemakers to Conquer the Past," *Time*, November 3, 2003, http://www.time.com/time/magazine/article/0, 9171,1125851,00.html.

3. See Robert Scales, Jr., *Certain Victory: The U.S. Army in the Gulf War* (Washington, DC: Office of the Chief of Staff, United States Army, 1993); Richard Swain, *"Lucky War": Third Army in Desert Storm* (Fort Leavenworth, KS: U.S. Army Command and General Staff College Press, 1994).

4. See Mark Bowden, *Black Hawk Down: A Story of Modern War* (New York: Penguin Putnam Inc., 2000).

5. See Wesley K. Clark, *Waging Modern War* (New York: Public Affairs, 2001), 420–21.

6. See Thomas Donnelly, *Operation Iraqi Freedom: A Strategic Assessment* (Washington, DC: American Enterprise Institute, 2004), 52–54; and Michael Gordon and Bernard Trainor, *Cobra II: The Inside Story of the Invasion and Occupation of Iraq* (New York: Pantheon Books, 2006), 497–500.

7. Donnelly, *Operation Iraqi Freedom*, 44.

8. David Zuchino, *Thunder Run* (New York: Atlantic Monthly Press, 2004), 71–78.

9. Quoted in Donnelly, *Operation Iraqi Freedom*, 80.

10. George Packer, "The Lesson of Tal Afar: Is it Too Late for the Administration to Correct its Course in Iraq–" *The New Yorker*, April 10, 2006, 60–61.

11. See Oliver Poole, "Iraqis in Former Rebel Strongholds Now Cheer American Soldiers," *Telegraph* (London), December 19, 2005.

12. Condoleezza Rice, testimony before the United States Senate Foreign Relations Committee, Washington, DC, October 19, 2005.

13. George W. Bush, speech before the City Club of Cleveland, Cleveland, OH, March 20, 2006.

14. This does not mean that there was, literally, one soldier helping out forty persons, considering that many in the Third ACR served noncombat functions.

15. David H. Petraeus, "Learning Counterinsurgency: Observations from Iraq," *Military Review*, January-February 2006, 8.

16. Hizbollah was officially founded in 1985, but the elements of which it was formed were actively fighting Israel in 1982.

17. Andrew Exum, *Hizballah at War: A Military Assessment* (Washington, DC: Washington Institute for Near East Policy, 2006), 2.

18. The Shebaa Farms area, currently occupied by the Israelis, is claimed by Israel, Lebanon, and Syria. The validity of their respective claims is a topic for another paper.

19. Exum, *Hizballah at War*, 3–4.

20. David Makovsky and Jeffrey White, *Lessons and Implications of the Israeli-Hizballah War: A Preliminary Assessment* (Washington, DC: Washington Institute for Near East Policy, 2006), 10.

21. Ibid., 9.

22. Anthony Cordesman, *Preliminary "Lessons" of the Israeli-Hezbollah War* (Washington, DC: Center for Strategic and International Studies, 2006), 14.

23. Sarah Kreps, *The 2006 Lebanon War: Lessons Learned* (Carlisle, PA: Parameters, U.S. Army War College, 2007), 79–81.

24. Exum, *Hizballah at War*, 5.

25. Ibid.

26. Ibid., 5, 11.

27. Ibid., 8.

28. Ibid., 12.

29. Makovsky and White, *Lessons and Implications of the Israeli-Hizballah War*, 41.

30. David Makovsky and Jeffrey White write: "Especially in the beginning, troops were committed in small numbers in a raiding role. When they encountered greater-than-expected opposition, they suffered casualties, leading to uncoordinated escalation in the form of rescue efforts, which also became entangled in the fighting and suffered casualties." Ibid., 52.

31. Exum, *Hizballah at War*, 14.

32. Most of those hits and penetrations were caused by ATGMs, while a minority was caused by large antitank mines. This discussion is concerned with the former, as the latter can be combated by existing mine-clearing techniques and, being immobile, are less capable of decisively altering a battle.

33. Although only five tanks were permanently destroyed, the others might still be considered operational "losses," as the fact that a tank can eventually be repaired offers little consolation to the wounded crew who have to bail out under fire or the commander and troops on the battlefield who were counting on it.

34. Exum, *Hizballah at War*, 5.

35. Ibid., 3.

36. Makovsky and White, *Lessons and Implications of the Israeli-Hizballah War*, 45.

37. Kreps, *The 2006 Lebanon War*, 76–77.

38. Makovsky and White, *Lessons and Implications of the Israeli-Hizballah War*, 35.

39. Exum, *Hizballah at War*, 10.

40. Ibid., 1.

41. Kreps, *The 2006 Lebanon War*, 79.

42. "Colin Powell Says Iraq in a 'Civil War,'" CNN.com, November 29, 2006, http://www.cnn.com/2006/POLITICS/11/29/powell.iraq/.

43. Thomas Ricks, "Situation Called Dire in West Iraq," *Washington Post*, September 11, 2006.

44. John F. Burns, "US Hopes Success in Anbar, Iraq Can Be Repeated," *International Herald Tribune*, July 8, 2007.

45. "Iraqi Casualties by Province," http://icasualties.org/oif/Provincemap.aspx, 2003-2007.

46. Matthew Millham, "Commander of the 1st Brigade Combat Team Credits Iraqis for Ramadi Turnaround," *Stars and Stripes*, March 6, 2007, http://stripes.com/article.asp–section=104&article=43037&archive=true.

47. See Jim Michaels, "An Army Colonel's Gamble Pays Off in Iraq," *USA Today Online*, April 30, 2007, http://www.usatoday.com/news/world/iraq/2007-04-30-ramadi-colonel_N.htm. The details in the rest of this paragraph are from Michaels.

48. Ibid.

49. United States Central Command, "Al Anbar Region Makes Steady Improvements," news release, February 11, 2007.

50. Sudarsan Raghavan, "Maliki, Petraeus Visit Insurgent Hotbed in Iraq," *Washington Post*, March 14, 2007.

51. United States Department of Defense, "U.S. General Sees Cause for Optimism in Anbar Province," American Foreign Press Services, March 30, 2007.

52. Moro is the name given by the Spanish in the sixteenth century to the Muslim minority concentrated in the southern regions of the largely Christian Philippines. Presently, there are approximately five million Moros.

53. Zachary Abuza, *Balik-Terrorism: The Return of the Abu Sayyaf* (Carlisle, PA: Strategic Studies Institute of the U.S. Army War College, 2005), 2.

54. Ibid., 4.

55. Ibid., 5.

56. There is some uncertainty as to whether al Qaeda completely withdrew its support or merely reduced it. In the early 2000s, civilian Filipino officials asserted that there was no evidence that al Qaeda was presently supporting ASG; military officials, on the other hand, asserted that al Qaeda was in fact supporting ASG. Whatever the details, al Qaeda's ties—if any—to ASG in the late 1990s and early 2000s were certainly of an order of magnitude smaller than they had been in the early-to-mid 1990s. See Larry Niksch, *Abu Sayyaf: Target of Philippine-U.S. Anti-Terrorism Cooperation* (Washington, DC: Congressional Research Service, 2002), 4; Abuza, *Balik-Terrorism*, 8.

57. Operation Enduring Freedom–Philippines officially began in early 2002, but it is worth noting that U.S. Pacific Command dispatched a Special Forces assessment team to begin collecting intelligence in October 2001. Gregory Wilson, "Anatomy of a Successful COIN Operation: OEF-Philippines and the Indirect Approach," *Military Review*, November/December 2006, 6.

58. Abuza, *Balik-Terrorism*, 43–44.

59. Wilson, "Anatomy of a Successful COIN Operation," 7.

60. Ibid.

61. Ibid.

62. Ibid.

63. Abuza, *Balik-Terrorism*, 29–32.

64. Ibid., 27–28.

65. Niksch, *Abu Sayyaf*, 6.

66. The Moro National Liberation Front signed a peace accord with the government in 1996 which granted limited autonomy to certain regions with a Muslim majority.

67. To put this in perspective, the MILF-controlled island of Mindanao is some eight hundred miles away, in the southern Philippines. This distance forces the Armed Forces of the Philippines to split its resources if it wishes to deal with both threats.

68. Niksch, *Abu Sayyaf*, 41.

Chapter 4: What Kind of Force?

1. Clifford Krauss, "Marine Leader Contritely Admits He Erred on 'Singles Only' Order, *New York Times*, August 13, 1993, http://query.nytimes.com/gst/fullpage.html–res=9F0CE7D9173EF930A2575BC0A965958260.

2. David Galula, *Counterinsurgency Warfare: Theory and Practice* (1964; repr., Westport, CT: Praeger Security International, 2006), 93.

3. Col. (P) Robert Brown, "The Agile Leader Mind-Set: Leveraging the Power of Modularity in Iraq," *Military Review*, July/August 2007, 33.

4. Quoted in Martin Libicki and Stuart Johnson, eds., *Dominant Battlespace Knowledge* (Washington, DC: National Defense University Press, 1995), i.

5. Ibid., iv.

6. Galula, *Counterinsurgency Warfare*, 101.

7. Brown, "The Agile Leader Mind-Set," 34.

8. Col. Ralph O. Baker, "HUMINT-Centric Operations: Developing Actionable Intelligence in the Urban Counterinsurgency Environment," *Military Review*, March/April 2007, 13.

9. Ibid.

10. Ibid., 16.

11. Lt. Gen. John F. Kimmons, "Transforming Army Intelligence," *Military Review*, November/December 2006, 69–72.

12. See Barak A. Salmoni, "Advances in Predeployment Culture Training: The U.S. Marine Corps Approach," *Military Review*, November/December 2006, 79–88.

13. Kimmons, "Transforming Army Intelligence," 72.

14. Bruce R. Pirnie, Alan J. Vick, Adam Grissom, Karl P. Mueller, and David T. Orletsky, *Beyond Close Air Support: Forging a New Air-Ground Partnership* (Santa Moncia, CA: RAND Corporation, 2005).

15. Ibid., 108.

16. Guy Raz, "Air Force Plays Smaller Role in Iraq," *Morning Edition*, National Public Radio, October 10, 2007, http://www.npr.org/templates/story/story.php–storyId=15126640.

17. Report of Second Battalion, Second Regiment Infantry Fires Support Element, quoted in James Untersheher, "The Case for Cannons: Success in Current Operations, New Technology Keep Artillery in the Fight," *Armed Forces Journal*, October 2007, 41.

18. Raz, "Air Force Plays Smaller Role."

19. Lt. Col. (P) Christopher Hickey, "Principles and Priorities in Training for Iraq," *Military Review*, March/April 2007, 22.

20. See Charles C. Krulak, "The Strategic Corporal: Leadership in the Three-Block War," *Marines Magazine*, January 1999, http://www.au.af.mil/au/awc/awcgate/usmc/strategic_corporal.htm; see also Nathaniel Fick, "Manning the Barricades," *New York Times*, March 13, 2005; Joel Garreau, "Point Men for a Revolution: Can the Marines Survive a Shift from Platoons to Networks–" *Washington Post*, March 6, 1999; James E. Szepesy, "The Strategic Corporal and the Emerging Battlefield: The Nexus Between the USMC's Three-Block War Concept and Network Centric Warfare," MA thesis, Tufts University, March 2005, http://fletcher.tufts.edu/research/2005/Szepesy.pdf.

21. Hickey, "Principles and Priorities in Training for Iraq," 23.

22. Brown, "The Agile Leader Mind-Set," 36–37.

23. Ibid., 26.

24. Ibid., 38–40.

25. Department of Defense, *Quadrennial Defense Review Report*, February 6, 2006, http://www.defenselink.mil/qdr/report/Report20060203.pdf.

26. Lt. Gen. David Barno, "Fighting the 'Other War': Counterinsurgency Strategy in Afghanistan, 2003–2005," *Military Review*, September/October 2007, 37.

27. Col. Frederick Kienle, "Creating an Iraqi Army from Scratch: Lessons for the Future," *National Security Outlook*, Washington, DC, American Enterprise Institute, May 2007, 1.

28. Ibid., 2.

29. See Kienle, "Creating an Iraqi Army from Scratch," for further elaboration; see also Lt. Col. John A. Nagl, *Institutionalizing Adaptation: It's Time for a Permanent Advisor Corps* (Washington, DC: Center for a New American Security, 2007).

30. For further elaboration, see Frederick W. Kagan, *No Middle Way: The Challenge of Exit Strategies from Iraq* (Washington, DC: American Enterprise Institute, 2007), http://www.aei.org/publications/pubID.26760/pub_details.asp.

31. Barno, "Fighting the 'Other War,'" 38.

32. Ibid., 39.

33. Captain Ryan T. Kranc, "Advising Indigenous Forces," *Small Wars Journal* 8 (May 2007), http://smallwarsjournal.com/blog/2007/03/advising-indigenous-forces/.

34. For a fuller description of this model, see Bob Killebrew, *The Left-Hand Side of the Spectrum: Ambassadors and Advisors in Future U.S. Strategy* (Washington, DC: Center for a New American Security, 2007).

35. Kranc, "Advising Indigenous Forces."

Chapter 5: Costs

1. Indeed, marine leaders have suggested such a division of labor (withdrawing marine units from Iraq and army units from Afghanistan, consolidating a single service in each country). While immediately impractical, the approach is a sound one and a useful basis for force planning.

2. Quoted in Vance Serchuk, "Ethiopia versus the Islamists: What the U.S. Military Has Been Up To in the Horn of Africa," *Weekly Standard*, January 15, 2007.

3. Adm. Gregory Johnson (USN, Ret.), "The 'Long War' Demands Proactive Engagement in Africa," *Africa Policy Forum*, Center for Strategic and International Studies, Washington, DC, October 2, 2006.

4. Government Accountability Office, *Force Structure: Actions Needed to Improve Estimates and Oversight of Costs for Transforming the Army to a Modular Force* (Washington, DC: GAO, 2005), 5–6.

5. Eric K. Shinseki, "Statement Before the Committee on Armed Services," Senate Armed Services Committee, Washington, DC, March 1, 2000.

6. See Col. Brian G. Watson, "Reshaping the Expeditionary Army to Win Decisively: The Case for Greater Stabilization Capacity in the Modular Force," *Carlisle Papers* (Carlisle, PA: Strategic Studies Institute, 2005), 4.

7. See Mike Mount, "Troops Put Thorny Questions to Rumsfeld," CNN, December 9, 2004, http://www.cnn.com/2004/WORLD/meast/12/08/rumsfled. troops/.

8. Joseph Biden, "Sen. Biden Again Demands Investigation into MRAP Delay: 'Pentagon Needs to Come Clean,'" August 27, 2007, http://biden. senate.gov/newsroom/details.cfm–id=279021.

9. See Lawrence J. Korb, Loren B. Thompson, and Caroline P. Wadhams, *Army Equipment After Iraq* (Washington, DC: Lexington Institute and Center for American Progress, 2006); and Lawrence J. Korb, Max A. Bergmann, and Loren B. Thompson, *Marine Corps Equipment After Iraq* (Washington, DC: Lexington Institute and Center for American Progress, 2006).

10. Quoted in James Crawley, "Military Gear in Iraq Wears Out Very Fast; Replacing It Is Likely to Take Years and Billions of Dollars," *Winston-Salem Journal*, November 8, 2005.

11. Andrew Feikert, "U.S. Army and Marine Corps Equipment Requirements: Background and Issues for Congress," Congressional Research Service, June 15, 2007, 14–15.

12. Andrew Krepinevich and Dakota Wood, *Of IEDs and MRAPs: Force Protection in Complex Irregular Operations* (Washington, DC: Center for Strategic and Budgetary Assessments, 2007), 6.

13. Quoted in "Gates Designates MRAP Pentagon's 'Highest Priority' Acquisition Program," *Inside Defense*, May 8, 2007.

14. Lt. Gen. Raymond Odierno, "Counterinsurgency Guidance," Headquarters Multinational Force–Iraq, Baghdad, June 2007, http://small-warsjournal.com/documents/mncicoinguide.pdf.

15. "The Lion in Winter: Government, Industry, and U.S. Naval Shipbuilding Challenges," *Defense Industry Daily*, April 12, 2006, http://www.defenseindustrydaily.com/the-lion-in-winter-government-industry-and-us-naval-shipbuilding-challenges-02136.

16. Gen. James T. Conway, "Statement before the Senate Armed Services Committee on Readiness," Senate Armed Services Committee, Washington, DC, February 15, 2007, 8.

17. Ibid., 9.

18. Korb, Bergmann, and Thompson, *Marine Corps Equipment After Iraq*, 14–15.

19. For a more complete description of the FCS network concept, see Program Manager, "14 + 1 + 1 Systems Overview," *Future Combat Systems*,

March 14, 2007, 5–9, http://64.233.169.104/search–q=cache:9ymRP5Zdh2QJ:www.army.mil/fcs/whitepaper/FCSwhitepaper07.pdf+FCS+network+concept+march+14,+2007&hl=en&ct=clnk&cd=3&gl=us.

20. "The Budget for Fiscal Year 2008," Department of Defense, Washington, DC, February 2007, 44–45, http://www.gpoaccess.gov/usbudget/fy08/index.html.

21. Author interviews with senior army force planning and budgeting officials, April 2007.

22. Congressional Budget Office, "CBO's Economic Forecasts for Calendar Years 2007 to 2017," undated document, http://www.cbo.gov/budget/data/econproj.pdf.

About the Authors

Thomas Donnelly is a resident fellow in defense and security policy studies at the American Enterprise Institute. He is the coeditor, with Gary J. Schmitt, of *Of Men and Materiel: The Crisis in Military Resources* (AEI Press, 2007), and the author of *The Military We Need* (AEI Press, 2005), *Operation Iraqi Freedom: A Strategic Assessment* (AEI Press, 2004), and several other books. From 1995 to 1999, he was policy group director and a professional staff member for the House Armed Services Committee. Mr. Donnelly has also served as a member of the U.S.-China Economic and Security Review Commission. He is a former editor of *Armed Forces Journal*, *Army Times*, and *Defense News*.

Frederick W. Kagan is a resident scholar in defense and security policy studies at the American Enterprise Institute. Previously an associate professor of military history at the U.S. Military Academy at West Point, he is the author of *The End of the Old Order: Napoleon and Europe, 1801–1805* (Da Capo, 2006) and coauthor of *While America Sleeps: Self-Delusion, Military Weakness, and the Threat to Peace Today* (St. Martin's Press, 2000). His most recent book is *Finding the Target: The Transformation of American Military Policy* (Encounter Books, 2006). Mr. Kagan is also the author of several reports by the Iraq Planning Group at AEI. A contributing editor of *The Weekly Standard*, he has written numerous articles on defense and foreign policy issues for *Foreign Affairs*, the *Wall Street Journal*, the *Washington Post*, the *Los Angeles Times*, *Policy Review*, *Commentary*, *Parameters*, and other periodicals.

Index

www.ingramcontent.com/pod-product-compliance
Lightning Source LLC
Jackson TN
JSHW011937131224
75386JS00041B/1428